Essential Equations for the Civil PE Exam Using the HP 33s

David V. Morse, PhD, PE
R. Wane Schneiter, PhD, PE, DEE
Norman R. Voigt, PE, PLS
Bruce A. Wolle, MSE, PE

Professional Publications, Inc.
Belmont, CA

How to Locate and Report Errata for This Book

At Professional Publications, we do our best to bring you error-free books. But when errors do occur, we want to make sure you can view corrections and report any potential errors you find, so the errors cause as little confusion as possible.

A current list of known errata and other updates for this book is available on the PPI website at **www.ppi2pass.com/errata**. We update the errata page as often as necessary, so check in regularly. You will also find instructions for submitting suspected errata. We are grateful to every reader who takes the time to help us improve the quality of our books by pointing out an error.

ESSENTIAL EQUATIONS FOR THE CIVIL PE EXAM USING THE HP 33s

Current printing of this edition: 1

Printing History

edition number	printing number	update
1	1	New book.

Printed in the United States of America

Professional Publications, Inc.
1250 Fifth Avenue, Belmont, CA 94002
(650) 593-9119
www.ppi2pass.com

Library of Congress Cataloging-in-Publication Data
Essential equations for the civil PE exam using the HP 33s /
David V. Morse... [et al.].
p. cm.
ISBN-10: 1-59126-057-4
ISBN-13: 978-1-59126-057-8
1. Engineering--Examinations--Study guides. 2. Engineering--Problems, exercises, etc. 3. Engineering--Certification--United States.
I. Morse, David V., 1953-.
TA159.E77 2006
620.0076--dc22

2005058625

TABLE OF CONTENTS

ABOUT THE AUTHORS

David V. Morse, PhD, PE (structural), holds a bachelor's degree from Pennsylvania State University and a master's degree and doctorate from Carnegie Mellon University, all in civil engineering. He is an adjunct professor of engineering with Pennsylvania State University, where he developed and has taught the structural segment of the Penn State civil PE review course for the past 15 years.

R. Wane Schneiter, PhD, PE, DEE (environmental and water resources), is the Benjamin H. Powell, Jr., '36 Professor of Engineering in the civil and environmental engineering department at the Virginia Military Institute, where he is responsible for the environmental engineering program. He has been an environmental consultant to many industrial, commercial, and governmental clients and non-governmental organizations.

Norman R. Voigt, PE, PLS (transportation), is a registered civil engineer and land surveyor. He obtained his bachelor's and master's degrees in civil engineering from the University of Pittsburgh, where he is now on the adjunct faculty. For more than 25 years he has taught sections of the civil PE review course given by Pennsylvania State University.

Bruce A. Wolle, MSE, PE (geotechnical), holds a bachelor's degree from New Mexico State University and a master's degree from Arizona State University, both in civil engineering. Licensed in 17 states, he works in geotechnical and materials engineering. He has six years' experience as supervisor and instructor for review and exam prep courses for the American Society of Civil Engineers, and is an advisor for PPI's Civil Passing Zone.

PREFACE

In this book we will show you how to use the hp 33s calculator to save time and improve your performance on the civil PE examination.

Faced with a geotechnical problem involving the active earth pressure coefficient, for example, you might ordinarily need a few minutes to calculate the value of

$$k_a = \frac{\sin^2(\theta + \phi)}{\sin^2\theta \sin(\theta - \delta)\left(1 + \sqrt{\frac{\sin(\phi + \delta)\sin(\phi - \beta)}{\sin(\theta - \delta)\sin(\theta + \beta)}}\right)^2}$$

But if you have this equation already stored in the hp 33s, you just enter the values of θ, ϕ, δ, and β, and the value of k_a is immediately returned.

Or suppose you're solving a problem involving the specific speed of a pump. The formula you need is

$$n_s = \frac{n\sqrt{Q}}{h_A^{0.75}}$$

But let's say that for this problem you need the value of h_A, not n_s. If you have this equation stored in your hp 33s, you don't need to rearrange the equation. In the time it takes you to enter the values of n_s, n, and Q, you'll have the value of h_A.

We chose 50 of the most useful equations for the civil PE exam, 10 in each of the five disciplines—environmental, geotechnical, structural, transportation, and water resources. Then we worked out the exact steps, keystroke by keystroke, that you need to store them in your hp 33s. And we've included 15 practice problems with step-by-step solutions to show how quickly and accurately you can solve problems using stored equations.

We hope you find this book useful in preparing for your PE exam.

ACKNOWLEDGMENTS

Thanks are due to Tony Miller, who made many contributions to this book, including working out most of the keystroke sequences; and to Lori Buschbaum, who checked each sequence of keys to make sure they led to the intended results.

Thanks as well to the PPI editorial and production staff members who worked on this book, including Cathy Schrott, production manager; Sarah Hubbard, editorial manager; Miriam Hanes, typesetter; and Scott Marley, project editor.

INTRODUCTION

HOW TO USE THIS BOOK

The hp 33s is a scientific calculator currently approved by the NCEES for use during the civil PE examination. This book will give you practice with the calculator's Equation mode and SOLVE function, concentrating on the types of equations and problems you are likely to encounter on the examination.

The book is in five parts. Part A, the largest, shows how to store 50 common equations in the hp 33s. Part B shows how to use these stored equations to solve 15 practice problems. Part C demonstrates the use of some preprogrammed functions of the hp 33s that can be useful on the exam. Part D is for those who prefer Algebraic mode to Reverse Polish Notation mode, and it covers the most important differences between these two modes. Finally, Part E is a quick reference section, which you may want to have handy on exam day. To make lookup faster, it contains just the information that will be useful during the exam.

A list of all 50 equations begins on page 2, organized by the discipline and session each equation is most likely to appear in. It will help you determine which equations you want to store. The steps for entering the first equation in Part A are explained in particular detail, so it's a good place to start.

The instructions in this book are generally in three columns. The left column gives the exact sequence of keystrokes to enter, and the right column contains a brief description of the purpose of each step. The middle column shows what appears on the hp 33s display after each step. The hp 33s display has two lines. Whenever only one line is shown for a step, it is the lower.

An equation stored in the hp 33s can only contain the letters A through Z as variables. This creates difficulties when the variables to be entered include Greek letters, names longer than one letter, or the same letter in upper- and lowercase or with different subscripts. In this book, a Greek letter is usually replaced with either the first letter of its English name (such as B for beta) or by a letter that resembles it (such as W for omega).

A common situation involves the same letter used with subscripts 1 and 2. In this book, the letter with subscript 2 drops the subscript, while the letter with subscript 1 drops the subscript and is shifted back in the alphabet one place. T_1 and T_2, for example, are represented by S and T, respectively. You can use different variables if you prefer.

At some point, the equation being entered may become longer than the display screen. Small left and right arrows (see Annunciators later in this introduction) will appear on the display, and the large silver key just below the display can be used to scroll left and right.

After the equation has been entered, pressing and holding the SHOW key will display a checksum and the length of the entered equation. Every sequence produces a unique checksum, so these numbers can be used to confirm that the equation has been entered correctly. If you use other variables than the ones in this book, you'll get a different checksum.

In the solution process, it sometimes happens that an equation has more than one possible root. This can happen, for example, when the unknown variable is squared in the equation. When the hp 33s can't solve an equation directly, it uses an iterative process, using one estimate of the answer to derive another, closer estimate. To direct the hp 33s to the right range of possible values, you may need to begin by entering a reasonable estimate for the variable. On the PE exam, a reasonable value can usually be determined from the answer choices given. More information about this can be found beginning on page 7-7 of the User's Manual.

The hp 33s has many times enough memory to store all the equations in this book. But the more equations you store, the longer it can take you to scroll through the list and find the one you want. You may find it best to store only those equations most likely to be useful to you. Taking a practice exam with the hp 33s can help you gauge how many equations you can comfortably handle on your list.

GETTING STARTED WITH THE HP 33s

If you are already familiar with the hp 33s, then skip the rest of this introduction, head to Part A, and start storing the equations you want to have handy. Otherwise, here is a brief overview focusing on what you will need to know to use the calculator during the exam. Other specific functions will be explained as they come up in the book. If you want to learn more, Appendix G in the User's Manual defines and explains the use of each and every key and function of the hp 33s.

In Equation mode, equations are entered in algebraic notation. The usual operator precedence is followed: first operators in parentheses, then exponents, multiplication and division, and finally addition and subtraction. Multiplication by adjacent parentheses is not recognized; the multiplication symbol must be used.

The layout of the hp 33s is shown here. The calculator is turned on by pressing [C] in the lower left corner. The calculator is turned off by pressing the purple [↱] key (located just above the [C] key) followed by [C].

Most keys have three functions. The main function is printed in white on the face of the key. These functions include the number keys, the operators [+], [−], [×], and [÷], and so on. To use a key's main function, press it without pressing a shift key first.

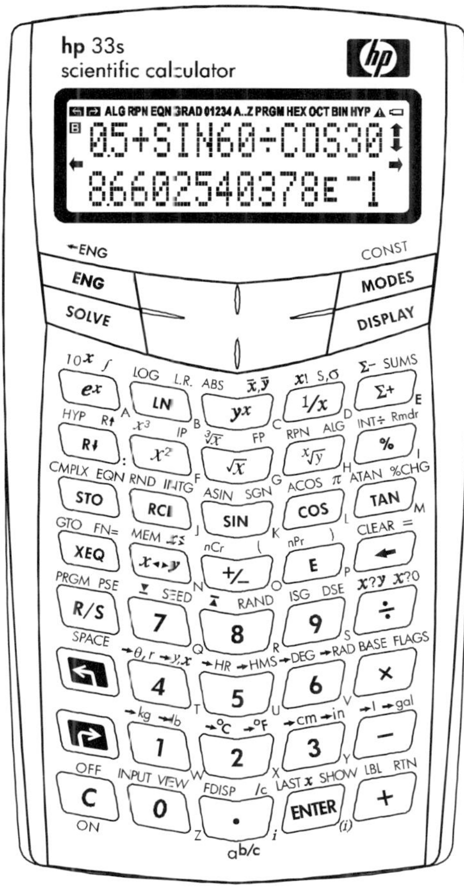

Shifted Keys

Above each key are printed two more functions, one in green and one in purple. To use one of these functions, first press the green or purple shift key (↰ or ↱), and then press the key for the function. Pressing a shift key a second time cancels the shift.

Some green and purple functions are grouped by field. Statistical functions are in the top row, trigonometric functions are in the third row, and probability functions are in the fourth.

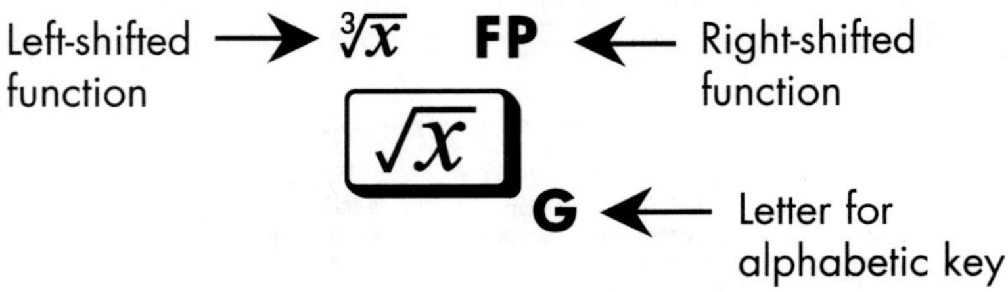

Alpha Keys

Some keys also have a letter of the alphabet printed to the lower right. Use these "alpha keys" whenever the hp 33s expects the name of a variable. Pressing RCL K, for example, puts the current value of variable K into the display; the RCL key signals that whatever key is pressed next will name a variable. When the alpha keys are active, the small **A..Z** annunciator appears at the top of the display.

Silver Keys

The large silver key is used to move through various lists or answers. This key can be thought of as having arrows at its indentations. Press on the top edge to scroll up, on the right edge to scroll to the right, and so on.

When there is information off the display that can be brought into view with the large silver key, small arrows will appear on the screen.

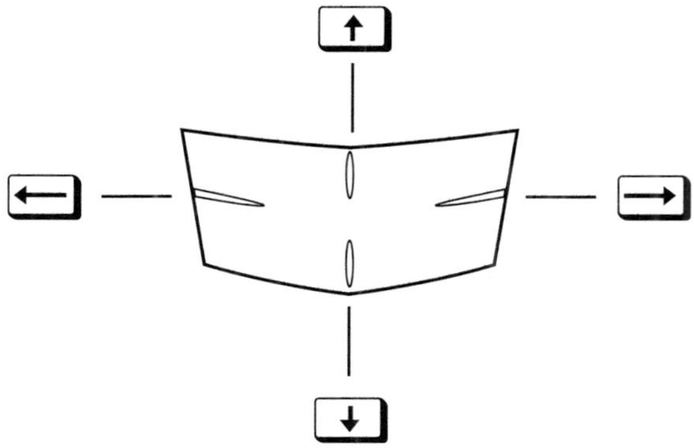

To the sides of the large silver key are four smaller silver keys. The pressure points on the keys are shown here.

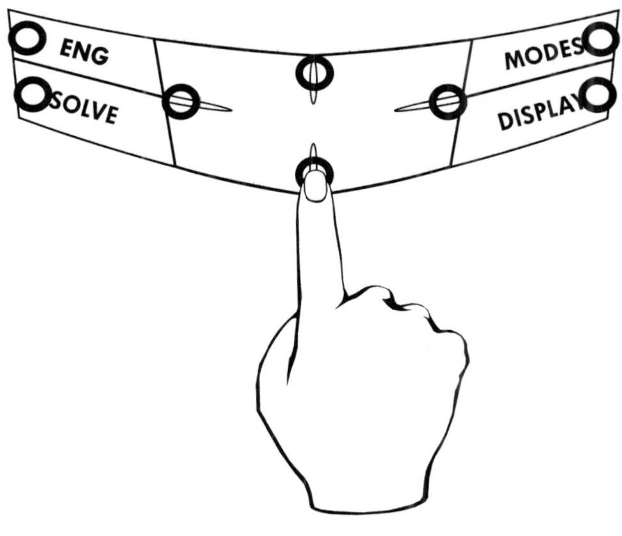

Annunciators

Small symbols called annunciators will appear on the display at times, usually to give information about the mode or status that the hp 33s is in. All the annunciators are shown here.

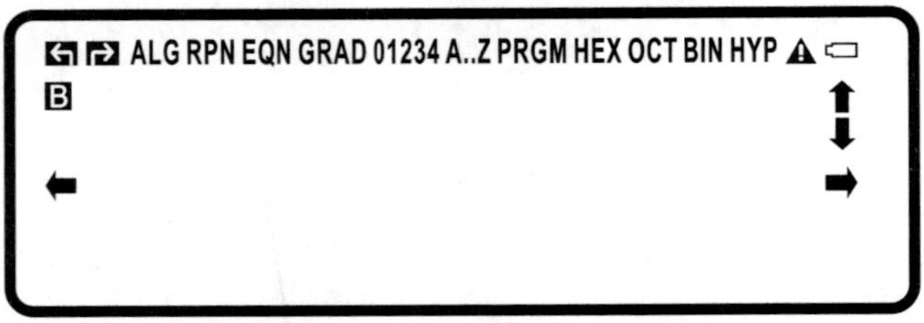

Correcting Errors

To correct errors, backspace with [←], or clear the number currently displayed with [C]. Pressing [↱] [CLEAR] and then selecting VARS will reset all variables to zero. These keys are explained in more detail in the User's Manual.

Equation Mode

To store a new question or solve one you've already stored, enter Equation mode by pressing [↱] [EQN].

RPN and ALG Modes

The calculator is normally in RPN mode. If it isn't, set it in this mode by pressing [↰] [RPN]. To select ALG mode, press [↱] [ALG].

EQUATION MODE VERSUS PROGRAMMING

These instructions show how to use the calculator's Equation mode, not its real programming capabilities. Programming is covered in the User's Manual, Chapters 12 to 17.

There are some advantages to writing programs for these equations instead of storing them in Equation mode. Programming lets you use word prompts for the variables, and allows you to retrieve an equation with a letter name instead of by scrolling through a list.

But against this are the disadvantages that programming on the hp 33s takes time and limits you to 26 equations. The aim of this book is to help you make the best use of your time in preparing for the civil PE exam, and using Equation mode is the way to do that.

REFERENCES

Problems and solutions in this book have been selected and adapted from the following books. All are from Professional Publications, Inc.

Lindeburg, Michael R. *Civil Engineering Reference Manual for the PE Exam.*

Schneiter, R. Wane. *Six-Minute Solutions for Civil PE Exam Environmental Problems.*

Schneiter, R. Wane. *Six-Minute Solutions for Civil PE Exam Problems: Water Resources.*

Subasic, Christine A. *Six-Minute Solutions for Civil PE Exam Structural Problems.*

Wolle, Bruce A. *Six-Minute Solutions for Civil PE Exam Geotechnical Problems.*

LIST OF EQUATIONS

This book contains keystroke-by-keystroke instructions for entering 50 equations into the hp 33s calculator. These equations are listed here arranged by the disciplines in which they are most likely to be used. (All examinees take the same morning session, and choose from among five possible afternoon sessions: environmental, geotechnical, structural, transportation, and water resources.) Equations that are likely to be useful on the morning (breadth) session are labeled AM. All other equations are likely to be useful on the afternoon (depth) session.

Which equations to store will depend on two things: which discipline your afternoon session will be in, and how many stored equations you can comfortably manage on your list. The hp 33s has enough memory to store not only all 50 equations but many hundreds more. But the more equations you store, the longer it will take to scroll through the list and find the one you want. So it is advantageous to store only those equations likely to be the most useful to you.

A sensible first step would be to store the AM equations for all five disciplines and the rest of the equations from your particular discipline. Then solve some practice problems with your hp 33s, or even take a complete sample examination. If you feel you can comfortably handle an even longer list, you might then add the equations from the other disciplines that will be represented on your afternoon session.

If your afternoon session will be in:	It will also include some questions in:
Environmental	Geotechnical and water resources
Geotechnical	Environmental, structural, and transportation
Structural	Geotechnical and transportation
Transportation	Geotechnical and water resources
Water resources	Environmental and geotechnical

A more detailed list of the subjects covered in each afternoon session can be found in PPI's *Civil Engineering Reference Manual* or on the NCEES website.

Environmental

1. Mass Balance — AM

$$C_f = \frac{C_1Q_1 + C_2Q_2}{Q_1 + Q_2}$$

2. Mean Cell Residence Time — AM

$$\theta_c = \frac{V_a X}{Q_e X_e + Q_w X_w}$$

3. Ultimate Biochemical Oxygen Demand — AM

$$\text{BOD}_t = \text{BOD}_u\left(1 - 10^{-K_d t}\right)$$

4. Population Equivalent

$$P_{\text{eq,1000s}} = \frac{\left(\text{BOD}_{\text{mg/L}}\right)\left(Q_{\text{ML/day}}\right)}{90\ \dfrac{\text{g}}{\text{person-day}}}$$

5. Time to Critical Point in Stream Purification

$$t_c = \left(\frac{1}{K_r - K_d}\right)\left(\log_{10}\left(\left(\frac{K_d\text{BOD}_u - K_r D_0 + K_d D_0}{K_d\text{BOD}_u}\right)\left(\frac{K_r}{K_d}\right)\right)\right)$$

6. Velocity of Settling

$$\text{v} = \sqrt{8k\left(\frac{gd_p}{f}\right)(\text{SG}_p - 1)}$$

7. Temperature Correction

$$K_{r,T_1} = K_{r,T_2}\theta_r^{T_1 - T_2}$$

8. Specific Substrate Utilization Rate

$$U = \frac{S_o - S}{\theta X}$$

9. Relationship Between Mean Cell Residence Time and Specific Utilization Rate

$$\frac{1}{\theta_c} = YU - k_d$$

10. Observed Yield

$$Y_{\text{obs}} = \frac{Y}{1 + k_d\theta_c}$$

Geotechnical

11. Mohr-Coulomb Equation AM

$$s = c + \sigma \tan \phi$$

12. Terzaghi-Meyerhof Equation for Continuous Footing AM

$$q_{\text{ult}} = \tfrac{1}{2}\gamma B N_\gamma + c N_c + (p_q + \gamma D_f) N_q$$

13. Primary Consolidation in Normally Consolidated Clay AM

$$S_{\text{primary}} = \frac{H C_c \log_{10}\left(\frac{p'_o + \Delta p'_v}{p'_o}\right)}{1 - e_o}$$

14. Active Earth Pressure Coefficient (Coulomb Equation) AM

$$k_a = \frac{\sin^2(\theta + \phi)}{\sin^2\theta \sin(\theta - \delta)\left(1 + \sqrt{\frac{\sin(\phi + \delta)\sin(\phi - \beta)}{\sin(\theta - \delta)\sin(\theta + \beta)}}\right)^2}$$

15. Piles: Point Capacity for Cohesive Soil

$$Q_p \approx 9 A_p c$$

16. Piles: Skin-Friction Capacity for Cohesive Soil

$$Q_f = p f_s L_e$$

17. Piles: Skin-Friction Capacity for Cohesive Soil (β-Method)

$$Q_f = p\beta\sigma' L_e$$

18. Increase in Applied Stress Below Loaded Rectangular Area

$$\Delta\sigma_z = q\frac{BL}{(B+z)(L+z)}$$

19. Safe Load for Double-Acting Steam Hammer

$$Q_{a,\text{lbf}} = \frac{2E_{\text{ft-lbf}}}{S_{\text{in}} + 0.1}$$

20. Time to Reach a Specific Vertical Consolidation

$$t = \frac{T_v H_d^2}{C_v}$$

Structural

21. Required Plate Thickness AM

$$t = 2c\sqrt{\frac{f_p}{F_y}}$$

(c = larger of m and n)

22. Deflection for Simply Supported Beam AM

$$\delta = \frac{5wL^4}{384EI}$$

23. Cross-Sectional Area of Circular Spiral Concrete Column AM

$$A_g = \frac{P_u}{\phi\beta\left(0.85f'_c(1-\rho_g) + \rho_g f_y\right)}$$

24. Shear Strength Provided by Concrete AM

$$V_c = 2\sqrt{f'_c}b_w d$$

25. Area of Steel in Reinforced Concrete Beam AM

$$A_s = \frac{M_u}{\phi f_y (d-\lambda)}$$

26. Distance From Neutral Axis to Bottom of Steel Beam

$$\bar{y}_b = \frac{A_s\left(\frac{d}{2}\right) + A_{\text{ctr}}\,(d + Y2)}{A_s + A_{\text{ctr}}}$$

27-28. Ordinates of Reinforced Concrete Interaction Diagram

$$x = \frac{\phi M_n}{A_g h}$$

$$y = \frac{\phi P_n}{A_g}$$

29. Development Length of Flexural Reinforcement

$$l_d = \frac{3 d_b f_y \beta \gamma}{50\sqrt{f_c'}}$$

30. Stability Interaction Criterion

$$\frac{f_a}{F_a} + \frac{C_{mx} f_{bx}}{\left(1 - \frac{f_a}{F_{ex}'}\right) F_{bx}} + \frac{C_{my} f_{by}}{\left(1 - \frac{f_a}{F_{ey}'}\right) F_{by}} \leq 1.0$$

Transportation

31. Middle Ordinate AM

$$M = R\left(1 - \cos\frac{I}{2}\right)$$

32. Elevation at Point on Curve AM

$$z_x = \left(\frac{R}{2}\right) x^2 + G_1 x + z_{\text{PVC}}$$

33-34. Stopping Sight Distance on Crest Vertical Curve AM

$$S < L: \quad L = \frac{AS^2}{200\left(\sqrt{h_1} + \sqrt{h_2}\right)^2}$$

$$S > L: \quad L = 2S - \frac{200\left(\sqrt{h_1} + \sqrt{h_2}\right)^2}{A}$$

35-36. Sight Distance on Sag Vertical Curve AM

$$S < L: \quad L = \frac{AS^2}{X + 3.5S}$$

$$S > L: \quad L = 2S - \left(\frac{X + 3.5S}{A}\right)$$

37. Heavy Vehicle Adjustment Factor AM

$$f_{HV} = \frac{1}{1 + P_T\left(E_T - 1\right) + P_R\left(E_R - 1\right)}$$

38. Constant Acceleration AM

$$s = s_0 + \mathrm{v}_0 t + \frac{a_0 t^2}{2}$$

39. Average Control Delay

$$d = 0.50C\left(\frac{\left(1 - \frac{g}{C}\right)^2}{1 - \left(\min\left(1, X\right)\left(\frac{g}{C}\right)\right)}\right)\left(\frac{\left(1 - P\right) f_{\mathrm{PA}}}{1 - \frac{g}{C}}\right) + 900T\left(\left(X - 1\right) + \sqrt{\left(X - 1\right)^2 + \left(\frac{8kIX}{cT}\right)}\right)$$

40. Free Flow Speed

$$\mathrm{FFS} = S_{\mathrm{FM}} + 0.00776\left(\frac{V_f}{f_{\mathrm{HV}}}\right)$$

Water Resources

41. Manning Equation AM

$$Q = \frac{K}{n} A R^{2/3} S^{1/2}$$

42. Energy Equation (Bernoulli Equation) AM

$$\frac{p_2}{\gamma} + z_2 + \frac{\mathrm{v}_2^2}{2g} + h_f = \frac{p_1}{\gamma} + z_1 + \frac{\mathrm{v}_1^2}{2g}$$

43. Darcy-Weisbach Equation AM

$$h_f = f\frac{L\mathrm{v}^2}{d2g}$$

44. Specific Speed of Pump

$$n_s = \frac{n\sqrt{Q}}{h_A^{0.75}}$$

45. Hydraulic Radius of Trapezoidal Channel

$$R = \frac{bd\sin\theta + d^2\cos\theta}{b\sin\theta + 2d}$$

46-47. Confined Stream in Pipe Bend

$$F_x = p_2A_2\cos\theta - p_1A_1 + \frac{\dot{m}\left(\mathrm{v}_2\cos\theta - \mathrm{v}_1\right)}{g_c}$$

$$F_y = \left(p_2A_2 + \frac{\dot{m}\mathrm{v}_2}{g_c}\right)\sin\theta$$

48. Conjugate Depths for Rectangular Channel

$$d_1 = -\frac{1}{2}d_2 + \sqrt{\frac{2\mathrm{v}_2^2 d_2}{g} + \frac{d_2^2}{4}}$$

$$d_2 = -\frac{1}{2}d_1 + \sqrt{\frac{2\mathrm{v}_1^2 d_1}{g} + \frac{d_1^2}{4}}$$

49. FAA Formula for Time to Concentration

$$t_c = \frac{(1.8)\,(1.1 - C)\,\sqrt{L}}{S^{1/3}}$$

50. Throat Velocity in Venturi Meter

$$\mathrm{v}_2 = \left(\frac{C_\mathrm{v}}{\sqrt{1-\beta^4}}\right)\sqrt{\frac{2g\left(\rho_m - \rho\right)h}{\rho}}$$

STORING EQUATIONS

A

ENVIRONMENTAL

Equation 1: Mass Balance

$$C_f = \frac{C_1Q_1 + C_2Q_2}{Q_1 + Q_2}$$

Entered as

`F=(B×P+C×Q)÷(P+Q)`

Discipline Environmental

Variables (See How to Use This Book for more information.)

$F = C_f$	final concentration	mg/gal	mg/L
$B = C_1$	concentration	mg/gal	mg/L
$C = C_2$	concentration	mg/gal	mg/L
$P = Q_1$	flow quantity	gal/day	L/d
$Q = Q_2$	flow quantity	gal/day	L/d

Notes This equation can be used to calculate other quantities than concentration, such as temperature, in a mixture of two flows.

Keys	Display	Description
[right-shift]	0.0000	Selects right-shift (purple) functions. Display will read 0.0000 if calculator is set to four decimal places and contains a value of zero. Continue in any case.

[EQN]	`EQN LIST TOP`	Selects Equation mode. **EQN** annunciator appears. If this is not first equation entered, other equations may be displayed. Continue in any case.
[RCL]	`▮`	Starting new equation causes rectangular cursor to appear. Pressing RCL causes **A..Z** annunciator to appears. Now you can enter a variable.
[F]	`F▮`	Enters variable *F*, representing final concentration (C_f).
[↱] [=]	`F=▮`	Enters equal sign.
[↱] [(]	`F=(▮`	Enters left parenthesis. Parentheses are used to change the order of operations, in this case to force the two additions to be calculated before the division.
[RCL] [B]	`F=(B▮`	*B* is concentration (C_1).
[×]	`F=(B×▮`	Enters multiplication sign.
[RCL] [P]	`F=(B×P▮`	*P* is flow quantity (Q_1).
[+]	`F=(B×P+▮`	Enters addition sign.
[RCL] [C]	`F=(B×P+C▮`	*C* is concentration (C_2).
[×] [RCL] [Q]	`F=(B×P+C×Q▮`	*Q* is flow quantity (Q_2).
[↱] [)]	`F=(B×P+C×Q)▮`	Enters right parenthesis.
[÷]	`F=(B×P+C×Q)÷▮`	Enters division sign.

Keys	Display	Description
[⮫] [(] [RCL] [P]	`P =(B×P+C×Q)÷(P▮`	Equation is now longer than the hp 33s display. For instructions on scrolling left and right, see How to Use This Book.
[+] [RCL] [Q] [⮫] [)]	`×P+C×Q)÷(P+Q)▮`	
[ENTER]	`P=(B×P+C×Q)÷(P`	Ends equation.
[⮫] [SHOW]	`CK=E72C` `LN=17`	Hold down SHOW to display checksum and length of equation. If numbers displayed match these, equation is correctly entered.

Equation 2: Mean Cell Residence Time

$$\theta_c = \frac{V_a X}{Q_e X_e + Q_w X_w}$$

Entered as

`T=V×X÷(P×E+Q×W)`

Discipline Environmental

Variables (See How to Use This Book for more information.)

$T = \theta_c$	mean cell residence time	day	d
$V = V_a$	volume of aeration tank	ft^3	m^3
$X = X$	mixed liquor volatile suspended solids	mg/L	mg/L
$P = Q_e$	flow rate, effluent	ft^3/day	m^3/d
$Q = Q_w$	flow rate, wasted	ft^3/day	m^3/d
$E = X_e$	mixed liquor volatile suspended solids, effluent	mg/L	mg/L
$W = X_w$	mixed liquor volatile suspended solids, wasted	mg/L	mg/L

Keys	Display	Description
[⮫] [EQN] [RCL] [T] [⮫] [=]	`T=▮`	Selects Equation mode and starts new equation. *T* is time (θ_c).
[RCL] [V]	`T=V▮`	*V* is tank volume (V_a).
[×] [RCL] [X]	`T=V×X▮`	*X* is suspended solids (X).

Keys	Display	Description
[÷] [↱] [(] [RCL] [P]	`T=V×X÷(P▮`	*P* is effluent flow rate (Q_e).
[×] [RCL] [E]	`T=V×X÷(P×E▮`	*E* is effluent solids (X_e).
[+] [RCL] [Q]	`T=V×X÷(P×E+Q▮`	*Q* is wasted flow rate (Q_w).
[×] [RCL] [W] [↱] [)]	`V×X÷(P×E+Q×W)▮`	*W* is wasted solids (X_w). Equation is now longer than the hp 33s display. For instructions on scrolling left and right, see How to Use This Book.
[ENTER]	`T=V×X÷(P×E+Q×W`	Ends equation.
[↱] [SHOW]	`CK=A904` `LN=15`	Hold down SHOW to display checksum and length of equation. If numbers displayed match these, equation is correctly entered.

Equation 3: Ultimate Biochemical Oxygen Demand

$$\text{BOD}_t = \text{BOD}_u \left(1 - 10^{-K_d t}\right)$$

Entered as

`T=U×(1-10^(-K×S))`

Discipline Environmental

Variables (See How to Use This Book for more information.)

$T = \text{BOD}_t$	biochemical oxygen demand at time t	mg/L
$U = \text{BOD}_u$	ultimate biochemical oxygen demand	mg/L
$K = K_d$	rate constant	d^{-1}
$S = t$	time	d

Keys	Display	Description
[↱] [EQN] [RCL] [T] [↱] [=]	`T=▮`	Selects Equation mode and starts new equation. *T* is biochemical oxygen demand (BOD_t).
[RCL] [U]	`T=U▮`	*U* is ultimate biochemical oxygen demand (BOD_u).

[×] [↱] [(] [1]	`T=U×( 1_`	Cursor changes for entering digits.
[−] [1] [0]	`T=U×(1- 10_`	
[yˣ] [↱] [(] [+/−]	`T=U×(1-10^(-▮`	Changes sign of number.
[RCL] [K]	`T=U×(1-10^(-K▮`	*K* is a rate constant (K_d).
[×] [RCL] [S]	`U×(1-10^(-K×S▮`	*S* is time (*t*). Equation is now longer than the hp 33s display. For instructions on scrolling left and right, see How to Use This Book.
[↱] [)] [↱] [)]	`(1-10^(-K×S))▮`	
[ENTER]	`T=U×(1-10^(-K×`	Ends equation.
[↱] [SHOW]	`CK=5209` `LN=17`	Hold down SHOW to display checksum and length of equation. If numbers displayed match these, equation is correctly entered.

Equation 4: Population Equivalent

$$P_{\text{eq,1000s}} = \frac{\left(\text{BOD}_{\text{mg/L}}\right)\left(Q_{\text{ML/day}}\right)}{90\ \dfrac{\text{g}}{\text{person-day}}}$$

Entered as

`P=B×Q÷90`

Discipline Environmental

Variables (See How to Use This Book for more information.)

$P = P_{\text{eq,1000s}}$	population in thousands	–
$B = \text{BOD}_{\text{mg/L}}$	biochemical oxygen demand	mg/L
$Q = Q_{\text{ML/day}}$	flow quantity	ML/d

Keys	Display	Description
[⮫] [EQN] [RCL] [P] [⮫] [=]	P=▮	Selects Equation mode and starts new equation. *P* is population ($P_{\text{eq,1000s}}$).
[RCL] [B]	P=B▮	*B* is biochemical oxygen demand ($\text{BOD}_{\text{mg/L}}$).
[×] [RCL] [Q]	P=B×Q▮	*Q* is flow quantity ($Q_{\text{ML/day}}$).
[÷] [9] [0]	P=B×Q÷ 90_	Cursor changes for entering digits.
[ENTER]	P=B×Q÷90	Ends equation.
[⮫] [SHOW]	CK=795B LN=8	Hold down SHOW to display checksum and length of equation. If numbers displayed match these, equation is correctly entered.

Equation 5: Time to Critical Point in Stream Purification

$$t_c = \left(\frac{1}{K_r - K_d}\right) \log_{10}\left(\left(\frac{K_d\text{BOD}_u - K_r D_0 + K_d D_0}{K_d\text{BOD}_u}\right)\left(\frac{K_r}{K_d}\right)\right)$$

Entered as

T=LOG((D×U−R×O+D×O)÷D÷U×R÷D)÷(R−D)

Discipline Environmental

Variables (See How to Use This Book for more information.)

$T = t_c$	time	d
$D = K_d$	rate constant, deoxygenated	d^{-1}
$U = \text{BOD}_u$	ultimate biochemical oxygen demand	mg/L
$R = K_r$	rate constant, reoxygenated	d^{-1}
$O = D_0$	initial oxygen deficit	mg/L

Keys	Display	Description
[⮫] [EQN] [RCL] [T] [⮫] [=]	T=▮	Selects Equation mode and starts new equation. *T* is time (t_c).

[↰] [LOG]	T=LOG(▮	Opens common logarithm function.
[↱] [(] [RCL] [D]	T=LOG((D▮	*D* is a rate constant (K_d).
[×] [RCL] [U]	T=LOG((D×U▮	*U* is ultimate biochemical oxygen demand (BOD_u).
[−] [RCL] [R]	T=LOG((D×U-R▮	*R* is a rate constant (K_r).
[×] [RCL] [O]	=LOG((D×U-R×O▮	*O* is initial oxygen deficit (D_0). Equation is now longer than the hp 33s display. For instructions on scrolling left and right, see How to Use This Book.
[+] [RCL] [D] [×] [RCL] [O] [↱] [)]	(D×U-R×O+D×O)▮	Closes parenthesis. Common logarithm function is still open.
[÷] [RCL] [D] [÷] [RCL] [U]	-R×O+D×O)÷D÷U▮	
[×] [RCL] [R] [÷] [RCL] [D] [↱] [)]	D×O)÷D÷U×R÷D)▮	Closes common logarithm function.
[÷] [↱] [(] [RCL] [R] [−] [RCL] [D] [↱] [)]	÷U×R÷D)÷(R-D)▮	
[ENTER]	T=LOG((D×U-R×O	Ends equation.
[↱] [SHOW]	CK=E615 LN=34	Hold down SHOW to display checksum and length of equation. If numbers displayed match these, equation is correctly entered.

Equation 6: Velocity of Settling

$$\mathrm{v} = \sqrt{8k\left(\frac{gd_p}{f}\right)(\mathrm{SG}_p - 1)}$$

Entered as

`V=SQRT(8×K×G×D÷F×(S-1))`

Discipline Environmental

Variables (See How to Use This Book for more information.)

$V = \mathrm{v}$	velocity	ft/sec	m/s
$K = k$	Camp constant	–	–
$G = g$	gravitational acceleration	32.2 ft/sec^2	9.81 m/s^2
$D = d_p$	diameter of particle	ft	m
$F = f$	Darcy friction factor	–	–
$S = \mathrm{SG}_p$	specific gravity	–	–

Keys	Display	Description
[↱] [EQN] [RCL] [V] [↱] [=]	`V=▮`	Selects Equation mode and starts new equation. *V* is velocity (v).
[√x]	`V=SQRT(▮`	Opens square root function.
[8]	`V=SQRT( 8_`	Cursor changes for entering digits.
[×] [RCL] [K]	`V=SQRT(8×K▮`	*K* is Camp constant (*k*).
[×] [RCL] [G]	`V=SQRT(8×K×G▮`	*G* is gravitational acceleration (*g*). When solving, enter 32.2 with U.S. units or 9.81 with SI units.
[×] [RCL] [D]	`=SQRT(8×K×G×D▮`	*D* is diameter of particle (d_p). Be careful to use consistent units. Equation is now longer than the hp 33s display. For instructions on scrolling left and right, see How to Use This Book.
[÷] [RCL] [F]	`QRT(8×K×G×D÷F▮`	*F* is Darcy friction factor (*f*).

Keys	Display	Description
[×] [↱] [(] [RCL] [S]	(8×K×G×D÷F×(S▮	*S* is specific gravity (SG_p).
[−] [1]	K×G×D÷F×(S- 1_	
[↱] [)] [↱] [)]	×G×D÷F×(S-1))▮	Closes square root function.
[ENTER]	V=SQRT(8×K×((G	Ends equation.
[↱] [SHOW]	CK=18A4 LN=23	Hold down SHOW to display checksum and length of equation. If numbers displayed match these, equation is correctly entered.

Equation 7: Temperature Correction

$K_{r,T_1} = K_{r,T_2}\theta_r^{T_1-T_2}$

Entered as

J=K×R^(S-T)

Discipline Environmental

Variables (See How to Use This Book for more information.)

$J = K_{r,T_1}$	rate constant	day^{-1}	d^{-1}
$K = K_{r,T_2}$	rate constant	day^{-1}	d^{-1}
$R = \theta_r$	temperature constant	–	–
$S = T_1$	temperature	°F	°C
$T = T_2$	temperature	°F	°C

Keys	Display	Description
[↱] [EQN] [RCL] [J] [↱] [=]	J=▮	Selects Equation mode and starts new equation. *J* is a rate constant (K_{r,T_1}).
[RCL] [K]	J=K▮	*K* is a rate constant ($K_{r,T2}$).
[×] [RCL] [R]	J=K×R▮	*R* is a temperature constant (θ_r).

Keys	Display	Description
[y^x]	J=K×R^▮	The operator ^ indicates an exponent. The expression r^{s-t} appears as R^($S-T$).
[↱] [(] [RCL] [S]	J=K×R^(S▮	S is temperature (T_1).
[−] [RCL] [T] [↱] [)]	J=K×R^(S−T)▮	T is temperature (T_2).
[ENTER]	J=K×R^(S−T)	Ends equation.
[↱] [SHOW]	CK=6797 LN=11	Hold down SHOW to display checksum and length of equation. If numbers displayed match these, equation is correctly entered.

Equation 8: Specific Substrate Utilization Rate

$$U = \frac{S_o - S}{\theta X}$$

Entered as

U=(O−S)÷T÷X

Discipline Environmental

Variables (See How to Use This Book for more information.)

$U = U$	specific substrate utilization	d^{-1}
$O = S_o$	growth-limited substrate concentration, influent	mg/L
$S = S$	growth-limited substrate concentration	mg/L
$T = \theta$	residence/detention time	d
$X = X$	mixed liquor volatile suspended solids	mg/L

Keys	Display	Description
[↱] [EQN] [RCL] [U] [↱] [=]	U=▮	Selects Equation mode and starts new equation. U is utilization.
[↱] [(] [RCL] [O]	U=(O▮	O is influent concentration (S_o).

[−] [RCL] [S]	U=(O−S▮	*S* is concentration.
[↱] [)] [÷] [RCL] [T]	U=(O−S)÷T▮	*T* is time (θ).
[÷] [RCL] [X]	U=(O−S)÷T÷X▮	*X* is suspended solids.
[ENTER]	U=(O−S)÷T÷X	Ends equation.
[↱] [SHOW]	CK=5CC1 LN=11	Hold down SHOW to display checksum and length of equation. If numbers displayed match these, equation is correctly entered.

Equation 9: Relationship Between Mean Cell Residence Time and Specific Utilization Rate

$$\frac{1}{\theta_c} = YU - k_d$$

Entered as

1÷T=Y×U−K

Discipline Environmental

Variables (See How to Use This Book for more information.)

$T = \theta_c$	mean cell residence time	d
$Y = Y$	yield coefficient	mg/mg
$U = U$	specific substrate utilization	d^{-1}
$K = k_d$	microorganism endogenous decay rate	d^{-1}

Keys	Display	Description
[↱] [EQN] [1]	1_	Selects Equation mode and starts new equation. Cursor changes for entering digits.
[÷] [RCL] [T]	1÷T▮	*T* is time (θ_c).
[↱] [=] [RCL] [Y]	1÷T=Y▮	*Y* is yield coefficient.
[×] [RCL] [U]	1÷T=Y×U▮	*U* is utilization.

[−] [RCL] [K]	`1÷T=Y×U-K▮`	*K* is decay rate (k_d).
[ENTER]	`1÷T=Y×U-K`	Ends equation.
[↱] [SHOW]	`CK=5ED8` `LN=9`	Hold down SHOW to display checksum and length of equation. If numbers displayed match these, equation is correctly entered.

Equation 10: Observed Yield

$$Y_{\text{obs}} = \frac{Y}{1 + k_d\theta_c}$$

Entered as

`O=Y÷(1+K×T)`

Discipline Environmental

Variables (See How to Use This Book for more information.)

$O = Y_{\text{obs}}$	observed yield coefficient	mg/mg
$Y = Y$	yield coefficient	mg/mg
$K = k_d$	microorganism endogenous decay rate	d^{-1}
$T = \theta_c$	cell residence time	d

Keys	Display	Description
[↱] [EQN] [RCL] [O] [↱] [=]	`O=▮`	Selects Equation mode and starts new equation. *O* is observed yield coefficient (Y_{obs}).
[RCL] [Y]	`O=Y▮`	*Y* is yield coefficient.
[÷] [↱] [(] [1]	`O=Y÷( 1_`	Cursor changes for entering digits.
[+] [RCL] [K]	`O=Y÷(1+K▮`	*K* is decay rate (k_d).
[×] [RCL] [T] [↱] [)]	`O=Y÷(1+K×T)▮`	*T* is time (θ_c).
[ENTER]	`O=Y÷(1+K×T)`	Ends equation.

[⮫] [SHOW]	CK=BF5A LN=11	Hold down SHOW to display checksum and length of equation. If numbers displayed match these, equation is correctly entered.

GEOTECHNICAL

Equation 11: Mohr-Coulomb Equation

$s = c + \sigma \tan \phi$

Entered as

S=C+N×TAN(P)

Discipline Geotechnical

Variables (See How to Use This Book for more information.)

$S = s$	shear strength	lbf/ft^2	Pa
$C = c$	cohesion	lbf/ft^2	Pa
$N = \sigma$	normal stress	lbf/ft^2	Pa
$P = \phi$	angle of internal friction	deg	deg

Keys	Display	Description
[⮫] [EQN] [RCL] [S] [⮫] [=]	S=■	Selects Equation mode and starts new equation. *S* is shear strength (*s*).
[RCL] [C]	S=C■	*C* is cohesion (*c*).
[+] [RCL] [N]	S=C+N■	*N* is normal stress (σ).
[×] [TAN]	S=C+N×TAN(■	Opens tangent function.
[RCL] [P]	S=C+N×TAN(P■	*P* is angle of internal friction (ϕ).
[⮫] [)]	S=C+N×TAN(P)■	Closes tangent function.
[ENTER]	S=C+N×TAN(P)	Ends equation.

[↱] [SHOW]	CK=6D58 LN=12	Hold down SHOW to display checksum and length of equation. If numbers displayed match these, equation is correctly entered.

Equation 12: Terzaghi-Meyerhof Equation for Continuous Footing

$q_{\text{ult}} = \frac{1}{2}\gamma B N_\gamma + c N_c + (p_q + \gamma D_f)\, N_q$

Entered as

Q=0.5×G×B×N+C×X+(P+G×D)×Y

Discipline Geotechnical

Variables (See How to Use This Book for more information.)

$Q = q_{\text{ult}}$	ultimate bearing capacity	lbf/ft^2	Pa
$G = \gamma$	specific weight	lbf/ft^3	N/m^3
$B = B$	width	ft	m
$N = N_\gamma$	bearing capacity factor, density	–	–
$C = c$	cohesion	lbf/ft^2	Pa
$X = N_c$	bearing capacity factor, cohesion	–	–
$P = p_q$	surcharge pressure	lbf/ft^2	Pa
$D = D_f$	depth of footing	ft	m
$Y = N_q$	bearing capacity factor, surcharge	–	–

Keys	Display	Description
[↱] [EQN] [RCL] [Q] [↱] [=]	Q=▮	Selects Equation mode and starts new equation. *Q* is ultimate bearing capacity (q_{ult}).
[·] [5]	Q= 0.5_	Cursor changes for entering digits.
[×] [RCL] [G]	Q=0.5×G▮	*G* is specific weight (γ).
[×] [RCL] [B]	Q=0.5×G×B▮	*B* is width.
[×] [RCL] [N]	Q=0.5×G×B×N▮	*N* is density bearing capacity factor (N_γ).
[+] [RCL] [C]	Q=0.5×G×B×N+C▮	*C* is cohesion (*c*).

[×] [RCL] [X]	`=0.5×G×B×N+C×X▮`	*X* is cohesion bearing capacity factor (N_c). Equation is now longer than the hp 33s display. For instructions on scrolling left and right, see How to Use This Book.
[+] [↱] [(] [RCL] [P]	`×G×B×N+C×X+(P▮`	*P* is pressure (p_q).
[+] [RCL] [G]	`×B×N+C×X+(P+G▮`	
[×] [RCL] [D]	`×N+C×X+(P−G×D▮`	*D* is footing depth (D_f).
[↱] [)] [×] [RCL] [Y]	`C×X+(P+G×D)×Y▮`	*Y* is surcharge bearing capacity factor (N_q).
[ENTER]	`Q=0.5×G×B×N+C×X`	Ends equation.
[↱] [SHOW]	`CK=A1DE` `LN=25`	Hold down SHOW to display checksum and length of equation. If numbers displayed match these, equation is correctly entered.

Equation 13: Primary Consolidation in Normally Consolidated Clay

$$S_{\text{primary}} = \frac{HC_c \log_{10}\left(\dfrac{p'_o + \Delta p'_v}{p'_o}\right)}{1 + e_o}$$

Entered as

`S=H×C×LOG((O+V)÷O)÷(1+E)`

Discipline Geotechnical

Variables (See How to Use This Book for more information.)

$S = S_{\text{primary}}$	settlement	ft	m
$H = H$	layer thickness or depth	ft	m
$C = C_c$	compression index	–	–
$O = p'_o$	original effective pressure	lbf/ft^2	Pa
$V = p'_v$	vertical effective pressure	lbf/ft^2	Pa
$E = e_o$	void ratio	–	–

Keys	Display	Description
[↱] [EQN] [RCL] [S] [↱] [=]	S=▮	Selects Equation mode and starts new equation. *S* is settlement ($S_{primary}$).
[RCL] [H]	S=H▮	*H* is layer thickness.
[×] [RCL] [C]	S=H×C▮	*C* is compression index (C_c).
[×] [↰] [LOG]	S=H×C×LOG(▮	Opens common logarithm function.
[↱] [(] [RCL] [O]	S=H×C×LOG((O▮	*O* is original effective pressure (p'_o).
[+] [RCL] [V]	=H×C×LOG((O+V▮	*V* is vertical effective pressure (p'_v). Equation is now longer than the hp 33s display. For instructions on scrolling left and right, see How to Use This Book.
[↱] [)] [÷] [RCL] [O] [↱] [)] [÷] [↱] [(] [1]	((O+V)÷O)÷(1_	Cursor changes for entering digits.
[+] [RCL] [E] [↱] [)]	O+V)÷O)÷(1+E)▮	*E* is void ratio (e_o).
[ENTER]	S=H×C×LOG((O+V	Ends equation.
[↱] [SHOW]	CK=FC37 LN=24	Hold down SHOW to display checksum and length of equation. If numbers displayed match these, equation is correctly entered.

Equation 14: Active Earth Pressure Coefficient (Coulomb Equation)

$$k_a = \frac{\sin^2(\theta+\phi)}{\sin^2\theta \sin(\theta-\delta)\left(1+\sqrt{\dfrac{\sin(\phi+\delta)\sin(\phi-\beta)}{\sin(\theta-\delta)\sin(\theta+\beta)}}\right)^2}$$

Entered as

```
K=SQ(SIN(T+P))÷SQ(SIN(T))÷SIN(T-D)÷SQ(1+SQRT(SIN(P+D)
    ×SIN(P-B)÷SIN(T-D)÷SIN(T+B)))
```

Discipline Geotechnical

Variables (See How to Use This Book for more information.)

$K = k_a$	active earth pressure coefficient	–	–
$T = \theta$	angle of resultant force	deg	deg
$P = \phi$	angle of internal friction	deg	deg
$D = \delta$	angle of external friction	deg	deg
$B = \beta$	slope of backfill	deg	deg

Keys	Display	Description
[↱] [EQN] [RCL] [K] [↱] [=]	K=▮	Selects Equation mode and starts new equation. *K* is active earth pressure coefficient (k_a).
[x^2]	K=SQ(▮	Opens square function.
[SIN]	K=SQ(SIN(▮	Opens sine function.
[RCL] [T]	K=SQ(SIN(T▮	*T* is angle of resultant force (θ).
[+] [RCL] [P]	K=SQ(SIN(T+P▮	*P* is angle of internal friction (ϕ).
[↱] [)]	K=SQ(SIN(T+P)▮	Closes sine function.
[↱] [)]	=SQ(SIN(T+P))▮	Closes square function. Equation is now longer than the hp 33s display. For instructions on scrolling left and right, see How to Use This Book.
[÷] [x^2] [SIN] [RCL] [T] [↱] [)] [↱] [)]	)÷(SQ(SIN(T))▮	

Keys	Display	Description
÷ SIN RCL T − RCL D ↱)	(T))÷SIN(T-D)▮	D is angle of external friction (δ).
÷ x^2 1	IN(T-D)÷SQ(1_	Cursor changes for entering digits.
+ √x SIN	Q(1+SQRT(SIN(▮	
RCL P + RCL D ↱)	SQRT(SIN(P+D)▮	
× SIN RCL P − RCL B ↱)	P+D)×SIN(P-B)▮	B is slope of backfill (β).
÷ SIN RCL T − RCL D ↱)	P-B)÷SIN(T-D)▮	
÷ SIN RCL T + RCL B ↱) ↱) ↱)	D)÷SIN(T+B)))▮	
ENTER	K=SQ(SIN(T+P))	Ends equation.
↱ SHOW	CK=D168 LN=82	Hold down SHOW to display checksum and length of equation. If numbers displayed match these, equation is correctly entered.

Equation 15: Piles: Point Capacity for Cohesive Soil

$Q_p \approx 9A_pc$

Entered as

Q=9×A×C

Discipline Geotechnical

Variables (See How to Use This Book for more information.)

$Q = Q_p$	bearing capacity, pile tip	lbf	N
$A = A_p$	area, pile tip	ft^2	m^2
$C = c$	cohesion	lbf/ft^2	Pa

Keys	Display	Description
[↱] [EQN] [RCL] [Q] [↱] [=]	`Q=■`	Selects Equation mode and starts new equation. *Q* is bearing capacity (Q_p).
[9]	`Q= 9_`	Cursor changes for entering digits.
[×] [RCL] [A]	`Q=9×A■`	*A* is area (A_p).
[×] [RCL] [C]	`Q=9×A×C■`	*C* is cohesion (*c*).
[ENTER]	`Q=9×A×C`	Ends equation.
[↱] [SHOW]	`CK=883B` `LN=7`	Hold down SHOW to display checksum and length of equation. If numbers displayed match these, equation is correctly entered.

Equation 16: Piles: Skin-Friction Capacity for Cohesive Soil

$$Q_f = pf_sL_e$$

Entered as

`Q=P×F×L`

Discipline Geotechnical

Variables (See How to Use This Book for more information.)

$Q = Q_f$	skin-friction capacity	lbf	N
$P = p$	perimeter	ft	m
$F = f_s$	skin friction factor	–	–
$L = L_e$	effective length	ft	m

Keys	Display	Description
[↱] [EQN] [RCL] [Q] [↱] [=]	Q=▮	Selects Equation mode and starts new equation. *Q* is skin-friction capacity (Q_f).
[RCL] [P]	Q=P▮	*P* is perimeter (*p*).
[×] [RCL] [F]	Q=P×F▮	*F* is skin friction factor (f_s).
[×] [RCL] [L]	Q=P×F×L▮	*L* is effective length (L_e).
[ENTER]	Q=P×F×L	Ends equation.
[↱] [SHOW]	CK=4DE4 LN=7	Hold down SHOW to display checksum and length of equation. If numbers displayed match these, equation is correctly entered.

Equation 17: Piles: Skin-Friction Capacity for Cohesive Soil (β-Method)

$$Q_f = p\beta\sigma' L_e$$

Entered as

Q=P×B×S×L

Discipline Geotechnical

Variables (See How to Use This Book for more information.)

$Q = Q_f$	skin-friction capacity	lbf	N
$P = p$	perimeter	ft	m
$B = \beta$	effective stress factor	–	–
$S = \sigma'$	earth pressure	–	–
$L = L_e$	effective length	ft	m

Keys	Display	Description
[↱] [EQN] [RCL] [Q] [↱] [=]	Q=▮	Selects Equation mode and starts new equation. *Q* is skin-friction capacity (Q_f).
[RCL] [P]	Q=P▮	*P* is perimeter (*p*).
[×] [RCL] [B]	Q=P×B▮	*B* is stress factor (β).

[×] [RCL] [S]	Q=P×B×S■	*S* is earth pressure (σ').
[×] [RCL] [L]	Q=P×B×S×L■	*L* is effective length (L_e).
[ENTER]	Q=P×B×S×L	Ends equation.
[↱] [SHOW]	CK=D4DC LN=9	Hold down SHOW to display checksum and length of equation. If numbers displayed match these, equation is correctly entered.

Equation 18: Increase in Applied Stress Below Loaded Rectangular Area

$$\Delta\sigma_z = q\frac{BL}{(B+z)(L+z)}$$

Entered as

S=Q×B×L÷(B+Z)÷(L+Z)

Discipline Geotechnical

Variables (See How to Use This Book for more information.)

$S = \sigma_z$	stress	lbf/ft^2	Pa
$Q = q$	uniform surcharge	lbf/ft^2	Pa
$B = B$	footing width	ft	m
$L = L$	distance	ft	m
$Z = z$	depth	ft	m

Keys	Display	Description
[↱] [EQN] [RCL] [S] [↱] [=]	S=■	Selects Equation mode and starts new equation. *S* is stress (σ_z).
[RCL] [Q]	S=Q■	*Q* is uniform surcharge (*q*).
[×] [RCL] [B]	S=Q×B■	*B* is footing width.
[×] [RCL] [L]	S=Q×B×L■	*L* is distance.
[÷] [↱] [(] [RCL] [B] [+] [RCL] [Z]	S=Q×B×L÷(B+Z■	*Z* is depth (*z*).

[↱] [)] [÷]	`=Q×B×L÷(B+Z)÷■`	Equation is now longer than the hp 33s display. For instructions on scrolling left and right, see How to Use This Book.
[↱] [(] [RCL] [L] [+] [RCL] [Z] [↱] [)]	`L÷(B+Z)÷(L+Z)■`	
[ENTER]	`S=Q×B×L÷(B+Z)÷`	Ends equation.
[↱] [SHOW]	`CK=F02E` `LN=19`	Hold down SHOW to display checksum and length of equation. If numbers displayed match these, equation is correctly entered.

Equation 19: Safe Load for Double-Acting Steam Hammer

$$Q_{a,\text{lbf}} = \frac{2E_{\text{ft-lbf}}}{S_{\text{in}} + 0.1}$$

Entered as

`Q=2×E÷(S+0.1)`

Discipline Geotechnical

Variables (See How to Use This Book for more information.)

$Q = Q_{a,\text{lbf}}$	allowable vertical load per pile	lbf	N
$E = E_{\text{ft-lbf}}$	energy per blow	ft-lbf	J
$S = S_{\text{in}}$	average penetration per blow	in	m

Keys	Display	Description
[↱] [EQN] [RCL] [Q] [↱] [=]	`Q=■`	Selects Equation mode and starts new equation. *Q* is load per pile ($Q_{a,\text{lbf}}$).
[2] [×] [RCL] [E]	`Q=2×E■`	*E* is energy per blow ($E_{\text{ft-lbf}}$).
[÷] [↱] [(] [RCL] [S]	`Q=2×E÷(S■`	*S* is penetration per blow (S_{in}).

[+] [.] [1]	`Q=2×E÷(S+ 0.1_`	Cursor changes for entering digits.
[↱] [)]	`Q=2×E÷(S+0.1)▮`	
[ENTER]	`Q=2×E÷(S+0.1)`	Ends equation.
[↱] [SHOW]	`CK=2756` `LN=13`	Hold down SHOW to display checksum and length of equation. If numbers displayed match these, equation is correctly entered.

Equation 20: Time to Reach Specific Vertical Consolidation

$$t = \frac{T_v H_d^2}{C_v}$$

Entered as

`T=V×SQ(H)÷C`

Discipline Geotechnical

Variables (See How to Use This Book for more information.)

$T = t$	time	sec	s
$V = T_v$	time factor	–	–
$H = H_d$	thickness of layer	ft	m
$C = C_v$	coefficient of consolidation	ft^2/sec	m^2/s

Notes H is full layer thickness for single drainage, half for double drainage. If the coefficient of consolidation is given in units of feet or meters per day, time will be in units of days. Take care to use consistent units.

Keys	**Display**	**Description**
[↱] [EQN] [RCL] [T] [↱] [=]	`T=▮`	Selects Equation mode and starts new equation. *T* is time (*t*).
[RCL] [V]	`T=V▮`	*V* is time factor (T_v).
[×] [x^2]	`T=V×SQ(▮`	Opens square function.
[RCL] [H] [↱] [)]	`T=V×SQ(H)▮`	*H* is layer thickness or depth (H_d). Closes square function.

[÷] [RCL] [C]	T=V×SQ(H)÷C▮	*C* is coefficient of consolidation (C_v).
[ENTER]	T=V×SQ(H)÷C	Ends equation.
[↱] [SHOW]	CK=EF39 LN=11	Hold down SHOW to display checksum and length of equation. If numbers displayed match these, equation is correctly entered.

STRUCTURAL

Equation 21: Required Plate Thickness

$$t = 2c\sqrt{\frac{f_p}{F_y}}$$

(c = larger of m and n)

Entered as

T=2×C×SQRT(P÷Y)

Discipline Structural

Variables (See How to Use This Book for more information.)

$T = t$	thickness	in
$C = c$	dimension	in
$P = f_p$	bearing stress	kips/in^2
$Y = F_y$	yield strength	kips/in^2

Keys	Display	Description
[↱] [EQN] [RCL] [T] [↱] [=]	T=▮	Selects Equation mode and starts new equation. *T* is thickness (t).
[2]	T= 2_	Cursor changes for entering digits.
[×] [RCL] [C]	T=2×C▮	*C* is the longer of dimensions *m* and *n* (c).
[×] [√x]	T=2×C×SQRT(▮	Opens square root function.

[RCL] [P]	T=2×C×SQRT(P■	*P* is bearing stress (f_p).
[÷] [RCL] [Y]	=2×C×SQRT(P÷Y■	*Y* is yield strength (F_y). Equation is now longer than the hp 33s display. For instructions on scrolling left and right, see How to Use This Book.
[↱] [)]	2×C×SQRT(P÷Y)■	Closes square root function.
[ENTER]	T=2×C×SQRT(P×Y	Ends equation.
[↱] [SHOW]	CK=F157 LN=15	Hold down SHOW to display checksum and length of equation. If numbers displayed match these, equation is correctly entered.

Equation 22: Deflection for Simply Supported Beam

$$\delta = \frac{5wL^4}{384EI}$$

Entered as

D=5×W×L^4÷384÷E÷I

Discipline Structural

Variables (See How to Use This Book for more information.)

$D = \delta$	deformation	in	m
$W = w$	load per unit length	lbf/in	N/m
$L = L$	length	in	m
$E = E$	modulus of elasticity	lbf/in^2	Pa
$I = I$	moment of inertia	in^4	m^4

Keys	Display	Description
[↱] [EQN] [RCL] [D] [↱] [=]	D=■	Selects Equation mode and starts new equation. *D* is deformation (δ).
[5]	D= 5_	Cursor changes for entering digits.
[×] [RCL] [W]	D=5×W■	*W* is load per unit length (w). Be careful to use consistent units.

[×] [RCL] [L]	`D=5×W×L■`	*L* is length.
[y^x]	`D=5×W×L^■`	Enters an exponent.
[4]	`D=5×W×L^ 4_`	L^4 is entered as *L*^4.
[÷] [3] [8] [4]	`=5×W×L^4÷ 384_`	Equation is now longer than the hp 33s display. For instructions on scrolling left and right, see How to Use This Book.
[÷] [RCL] [E]	`5×W×L^4÷384÷E■`	*E* is modulus of elasticity.
[÷] [RCL] [I]	`W×L^4÷384÷E÷I■`	*I* is moment of inertia.
[ENTER]	`D=5×W×L^4÷(384`	Ends equation.
[↱] [SHOW]	`CK=3ACD` `LN=17`	Hold down SHOW to display checksum and length of equation. If numbers displayed match these, equation is correctly entered.

Equation 23: Cross-Sectional Area of Circular Spiral Concrete Column

$$A_g = \frac{P_u}{\phi\beta\left(0.85f'_c\left(1-\rho_g\right)+\rho_g f_y\right)}$$

Entered as

`A=U÷P÷B÷(0.85×C×(1-G)+G×Y)`

Discipline Structural

Variables (See How to Use This Book for more information.)

$A = A_g$	gross area	in^2	m^2
$U = P_u$	factored axial load	lbf	N
$P = \phi$	strength reduction factor	–	–
$B = \beta$	column strength factor	–	–
$C = f'_c$	compressive strength	lbf/in^2	Pa
$G = p_g$	longitudinal reinforcement ratio	–	–
$Y = f_y$	yield stress	lbf/in^2	Pa

Keys	Display	Description
[↱] [EQN] [RCL] [A] [↱] [=]	`A=▮`	Selects Equation mode and starts new equation. *A* is area (A_g).
[RCL] [U]	`A=U▮`	*U* is factored axial load (P_u).
[÷] [RCL] [P]	`A=U÷P▮`	*P* is strength reduction factor (ϕ).
[÷] [RCL] [B]	`A=U÷P÷B▮`	*B* is column strength factor (β).
[÷] [↱] [(] [·] [8] [5]	`A=U÷P÷B÷( 0.85_`	Cursor changes for entering digits.
[×] [RCL] [C]	`=U÷P÷B÷(0.85×C▮`	*C* is compressive strength (f'_c). Equation is now longer than the hp 33s display. For instructions on scrolling left and right, see How to Use This Book.
[×] [↱] [(] [1] [−] [RCL] [G]	`B÷(0.85×C×(1-G▮`	*G* is longitudinal reinforcement ratio (p_g).
[↱] [)] [+] [RCL] [G] [×] [RCL] [Y] [↱] [)]	`×C×(1-G)+G×Y)▮`	*Y* is yield stress (f_y).
[ENTER]	`A=U÷P÷B÷(0.85×`	Ends equation.
[↱] [SHOW]	`CK=3F28` `LN=26`	Hold down SHOW to display checksum and length of equation. If numbers displayed match these, equation is correctly entered.

Equation 24: Shear Strength Provided by Concrete

$V_c = 2\sqrt{f'_c} b_w d$

Entered as

V=2×SQRT(F)×B×D

Discipline Structural

Variables (See How to Use This Book for more information.)

$V = V_c$	shear strength, concrete	lbf
$F = f'_c$	compressive strength	lbf/in^2
$B = b_w$	width of compressive member, web	in
$D = d$	beam depth (to tension steel centroid)	in

Notes This equation is used only with U.S. units.

Keys	Display	Description
[↱] [EQN] [RCL] [V] [↱] [=]	V=▮	Selects Equation mode and starts new equation. *V* is shear strength (V_c).
[2]	V= 2_	Cursor changes for entering digits.
[×] [√x]	V=2×SQRT(▮	Opens square function.
[RCL] [F]	V=2×SQRT(F▮	*F* is compressive strength (f'_c).
[↱] [)]	V=2×SQRT(F)▮	Closes square function.
[×] [RCL] [B]	V=2×SQRT(F)×B▮	*B* is width (b_w).
[×] [RCL] [D]	2×SQRT(F)×B×D▮	*D* is beam depth (*d*). Equation is now longer than the hp 33s display. For instructions on scrolling left and right, see How to Use This Book.
[ENTER]	V=2×SQRT(F)×B×	Ends equation.
[↱] [SHOW]	CK=FBF9 LN=15	Hold down SHOW to display checksum and length of equation. If numbers displayed match these, equation is correctly entered.

Equation 25: Area of Steel in Reinforced Concrete Beam

$$A_s = \frac{M_u}{\phi f_y (d - \lambda)}$$

Entered as `A=M÷P÷F÷(D-L)`

Discipline Structural

Variables (See How to Use This Book for more information.)

$A = A_s$	area	in^2
$M = M_u$	ultimate moment	in-kips
$P = \phi$	strength reduction factor	–
$F = f_y$	yield stress	ksi
$D = d$	depth	in
$L = \lambda$	distance from centroid of compressed area to extreme compression fiber	in

Keys	Display	Description
[↱] [EQN] [RCL] [A] [↱] [=]	`A=■`	Selects Equation mode and starts new equation. *A* is area (A_s).
[RCL] [M]	`A=M■`	*M* is ultimate moment (M_u).
[÷] [RCL] [P]	`A=M÷P■`	*P* is strength reduction factor (ϕ).
[÷] [RCL] [F]	`A=M÷P÷F■`	*F* is yield stress (f_y).
[÷] [↱] [(] [RCL] [D]	`A=M÷P÷F÷(D■`	*D* is depth (d).
[−] [RCL] [L] [↱] [)]	`A=M÷P÷F÷(D-L)■`	*L* is distance from centroid to extreme compression fiber (λ).
[ENTER]	`A=M÷P÷F÷(D-L)`	Ends equation.
[↱] [SHOW]	`CK=7BDB` `LN=13`	Hold down SHOW to display checksum and length of equation. If numbers displayed match these, equation is correctly entered.

Equation 26: Distance From Neutral Axis to Bottom of Steel Beam

$$\bar{y}_b = \frac{A_s\left(\dfrac{d}{2}\right) + A_{\text{ctr}}\,(d + Y2)}{A_s + A_{\text{ctr}}}$$

Entered as

`B=(S×D÷2+C×(D+Y))÷(S+C)`

Discipline Structural

Variables (See How to Use This Book for more information.)

$B = \bar{y}_b$	distance from neutral axis of composite beam to bottom of steel beam under full composite action	in
$S = A_s$	area of steel beam	in^2
$D = d$	depth of steel beam	in
$C = A_{\text{ctr}}$	area of transformed concrete	in^2
$Y = Y2$	distance from top of steel beam to centroid of concrete compressive area	in

Keys	Display	Description
[↱] [EQN] [RCL] [B] [↱] [=]	`B=▮`	Selects Equation mode and starts new equation. *B* is distance from neutral axis to bottom of beam ($\bar{y}_b$).
[↱] [(] [RCL] [S]	`B=(S▮`	*S* is area of steel beam (A_s).
[×] [RCL] [D]	`B=(S×D▮`	*D* is depth of steel beam (*d*).
[÷] [2]	`B=(S×D÷ 2_`	Cursor changes for entering digits.
[+] [RCL] [C]	`B=(S×D÷2+C▮`	*C* is area of concrete (A_{ctr}).
[×] [↱] [(] [RCL] [D] [+] [RCL] [Y]	`(S×D÷2+C×(D+Y▮`	*Y* is distance from top of beam to centroid of concrete (*Y2*). Equation is now longer than the hp 33s display. For instructions on scrolling left and right, see How to Use This Book.
[↱] [)] [↱] [)] [÷] [↱] [(] [RCL] [S]	`2+C×(D+Y))÷(S▮`	

[+] [RCL] [C] [↱] [)]	`×(D+Y))÷(S−C)▮`	
[ENTER]	`B=(S×D÷2+C×(D+`	Ends equation.
[↱] [SHOW]	`CK=D7EA` `LN=23`	Hold down SHOW to display checksum and length of equation. If numbers displayed match these, equation is correctly entered.

Equations 27-28: Ordinates of Reinforced Concrete Interaction Diagram

$$x = \frac{\phi M_n}{A_g h}$$

$$y = \frac{\phi P_n}{A_g}$$

Entered as

`X=P×M÷A÷H`

`Y=P×N÷A`

Discipline Structural

Variables (See How to Use This Book for more information.)

$X = x$	value on x-axis	ksi
$P = \phi$	strength reduction factor	–
$M = M_n$	nominal moment	in-kips
$A = A_g$	gross area	in^2
$H = h$	effective height	in
$Y = y$	value on y-axis	ksi
$N = P_n$	nominal axial load	kips

Notes These equations are used to find quickly the x- and y-values to use with a reinforced concrete interaction diagram. If the two equations are used one right after the other, the values for ϕ and A_g do not have to be keyed in again the second time.

Keys	Display	Description
[↱] [EQN] [RCL] [X] [↱] [=]	`X=▮`	Selects Equation mode and starts new equation. *X* is value on *x*-axis.

[RCL] [P]	X=P■	*P* is strength reduction factor (ϕ).
[×] [RCL] [M]	X=P×M■	*M* is nominal moment (M_n).
[÷] [RCL] [A]	X=P×M÷A■	*A* is gross area (A_g).
[÷] [RCL] [H]	X=P×M÷A÷H■	*H* is effective height (h).
[ENTER]	X=P×M÷A÷H	Ends equation.
[↱] [SHOW]	CK=D0F1 LN=9	Hold down SHOW to display checksum and length of equation. If numbers displayed match these, equation is correctly entered.

Keys	Display	Description
[↱] [EQN] [RCL] [Y] [↱] [=]	Y=■	Selects Equation mode and starts new equation. *Y* is value on *y*-axis.
[RCL] [P]	Y=P■	
[×] [RCL] [N]	Y=P×N■	*N* is nominal axial load (P_n).
[÷] [RCL] [A]	Y=P×N÷A■	
[ENTER]	Y=P×N÷A	Ends equation.
[↱] [SHOW]	CK=F316 LN=7	Hold down SHOW to display checksum and length of equation. If numbers displayed match these, equation is correctly entered.

Equation 29: Development Length of Flexural Reinforcement

$$l_d = \frac{3d_b f_y \beta\gamma}{50\sqrt{f_c'}}$$

Entered as

L=3×D×Y×B×G÷50÷SQRT(C)

Discipline Structural

Variables (See How to Use This Book for more information.)

$L = l_d$	development length	in
$D = d_b$	diameter of bar	in
$Y = f_y$	yield strength of steel	lbf/in^2
$B = \beta$	epoxy coating factor	–
$G = \gamma$	size factor	–
$C = f'_c$	compressive strength of concrete	lbf/in^2

Notes The values of B (β) and G (γ) are:

$B = 1.5$	epoxy-coated, clear spacing $\leq 6d_b$
$B = 1.2$	for all other epoxy-coated bars
$B = 1.0$	uncoated bars
$G = 0.8$	no. 6 bars and smaller
$G = 1.0$	no. 7 bars and larger

Keys	Display	Description
[↱] [EQN] [RCL] [L] [↱] [=]	`L=▮`	Selects Equation mode and starts new equation. *L* is development length (l_d).
[3]	`L= 3_`	Cursor changes for entering digits.
[×] [RCL] [D]	`L=3×D▮`	*D* is diameter (d_b).
[×] [RCL] [Y]	`L=3×D×Y▮`	*Y* is yield strength (f_y).
[×] [RCL] [B]	`L=3×D×Y×B▮`	*B* is epoxy coating factor (β).
[×] [RCL] [G]	`L=3×D×Y×B×G▮`	*G* is specific weight (γ).
[÷] [5] [0]	`3×D×Y×B×G÷ 50_`	Equation is now longer than the hp 33s display. For instructions on scrolling left and right, see How to Use This Book.
[÷] [√x]	`×B×G÷50÷SQRT(▮`	Opens square root function.
[RCL] [C]	`B×G÷50÷SQRT(C▮`	*C* is compressive strength (f'_c).
[↱] [)]	`×G÷50÷SQRT(C)▮`	Closes square root function.
[ENTER]	`L=3×D×Y×B×G÷50`	Ends equation.

[↱] [SHOW]	CK=B61E LN=22	Hold down SHOW to display checksum and length of equation. If numbers displayed match these, equation is correctly entered.

Equation 30: Stability Interaction Criterion

$$\frac{f_a}{F_a} + \frac{C_{mx} f_{bx}}{\left(1 - \frac{f_a}{F'_{ex}}\right) F_{bx}} + \frac{C_{my} f_{by}}{\left(1 - \frac{f_a}{F'_{ey}}\right) F_{by}} \leq 1.0$$

Entered as

C=A÷B+M×R÷(1-A÷X)÷E+N×S÷(1-A÷Y)÷F

Discipline Structural

Variables (See How to Use This Book for more information.)

C	stability interaction criterion	–
$A = A_a$	allowable computed stress	kips/in^2
$B = F_a$	allowable stress	kips/in^2
$M = C_{mx}$	coefficient in interaction equation, strong axis	–
$R = f_{bx}$	computed stress, strong axis	kips/in^2
$X = F'_{ex}$	Euler stress divided by factor of safety, strong axis	kips/in^2
$E = F_{bx}$	allowable stress, strong axis	kips/in^2
$N = C_{my}$	coefficient in interaction equation, weak axis	–
$S = f_{by}$	computed stress, weak axis	kips/in^2
$Y = F'_{ey}$	Euler stress divided by factor of safety, weak axis	kips/in^2
$F = F_{by}$	allowable stress, weak axis	kips/in^2

Keys	Display	Description
[↱] [EQN] [RCL] [C] [↱] [=]	C=▮	Selects Equation mode and starts new equation. *C* is stability interaction criterion. If $C \leq 1.0$, then critical buckling stresses are assumed not to occur. If $C > 1.0$, then critical buckling stresses are assumed to occur.
[RCL] [A]	C=A▮	*A* is allowable computed stress (A_a).
[÷] [RCL] [B]	C=A÷B▮	*B* is allowable stress (F_a).

Keys	Display	Description
[+] [RCL] [M]	`C=A÷B+M▮`	*M* is coefficient in interaction equation, strong axis (C_{mx}).
[×] [RCL] [R]	`C=A÷B+M×R▮`	*N* is coefficient in interaction equation, weak axis (C_{my}).
[÷] [↱] [(] [1]	`C=A÷B+M×R÷( 1_`	Cursor changes for entering digits.
[−] [RCL] [A] [÷] [RCL] [X]	`÷B+M×R÷(1-A÷X▮`	*X* is Euler stress divided by factor of safety, strong axis (F'_{ex}). Equation is now longer than the hp 33s display. For instructions on scrolling left and right, see How to Use This Book.
[↱] [)] [÷] [RCL] [E]	`M×R÷(1-A÷X)÷E▮`	*E* is allowable stress, strong axis (F_{bx}).
[+] [RCL] [N]	`R÷(1-A÷X)÷E+N▮`	*N* is coefficient in interaction equation, weak axis (C_{my}).
[×] [RCL] [S]	`(1-A÷X)÷E+N×S▮`	*S* is computed stress, weak axis (f_{by}).
[÷] [↱] [(] [1] [−] [RCL] [A] [÷] [RCL] [Y]	`÷E+N×S÷(1-A÷Y▮`	*Y* is Euler stress divided by factor of safety, weak axis (F'_{ey}).
[↱] [)] [÷] [RCL] [F]	`N×S÷(1-A÷Y)÷F▮`	*F* is allowable stress, weak axis (F_{by}).
[ENTER]	`C=A÷B+M×R÷(1-A`	Ends equation.
[↱] [SHOW]	`CK=953F` `LN=33`	Hold down SHOW to display checksum and length of equation. If numbers displayed match these, equation is correctly entered.

TRANSPORTATION

Equation 31: Middle Ordinate

$$M = R\left(1 - \cos\frac{I}{2}\right)$$

Entered as

M=R×(1-COS(I÷2))

Discipline Transportation

Variables (See How to Use This Book for more information.)

$M = M$	middle ordinate	ft	m
$R = R$	curve radius	ft	m
$I = I$	interior angle	deg	deg

Keys	Display	Description
[↱] [EQN] [RCL] [M] [↱] [=]	M=▮	Selects Equation mode and starts new equation. *M* is middle ordinate.
[RCL] [R]	M=R▮	*R* is curve radius.
[×] [↱] [(] [1]	M=R×(1_	Cursor changes for entering digits.
[−] [COS]	M=R×(1-COS(▮	Opens cosine function.
[RCL] [I]	M=R×(1-COS(I▮	*I* is interior angle.
[÷] [2]	R×(1-COS(I÷ 2_	Equation is now longer than the hp 33s display. For instructions on scrolling left and right, see How to Use This Book.
[↱] [)] [↱] [)]	×(1-COS(I÷2))▮	Closes cosine function.
[ENTER]	M=R×(1-COS(I÷2	Ends equation.

Keys	Display	Description
[⮳] [SHOW]	CK=54AA LN=16	Hold down SHOW to display checksum and length of equation. If numbers displayed match these, equation is correctly entered.

Equation 32: Elevation at Point on Curve

$$z_x = \left(\frac{R}{2}\right) x^2 + G_1 x + z_{\text{BVC}}$$

Entered as

Z=R÷2×SQ(X)+G×X+B

Discipline Transportation

Variables (See How to Use This Book for more information.)

$Z = z_x$	elevation at point on curve	ft
$R = R$	rate of change	ft/sta^2
$X = x$	distance from beginning of vertical curve to point	sta
$G = G_1$	initial grade	ft/sta
$B = z_{\text{BVC}}$	elevation at beginning of vertical curve	ft

Keys	Display	Description
[⮳] [EQN] [RCL] [Z] [⮳] [=]	Z=▮	Selects Equation mode and starts new equation. *Z* is elevation at point on vertical curve (z_x).
[RCL] [R]	Z=R▮	*R* is curve radius.
[÷] [2]	Z=R÷ 2_	Cursor changes for entering digits.
[×] [x^2]	Z=R÷2×SQ(▮	Opens square function.
[RCL] [X]	Z=R÷2×SQ(X▮	*X* is distance from beginning of vertical curve to point (x).
[⮳] [)]	Z=R÷2×SQ(X)▮	Closes square function.
[+] [RCL] [G]	Z=R÷2×SQ(X)+G▮	*G* is grade (G_1).

Keys	Display	Description
[×] [RCL] [X]	`R÷2×SQ(X)+G×X▮`	Equation is now longer than the hp 33s display. For instructions on scrolling left and right, see How to Use This Book.
[+] [RCL] [B]	`2×SQ(X)+G×X+B▮`	*B* is elevation at beginning of vertical curve (z_{BVC}).
[ENTER]	`Z=R÷2×SQ(X)+G▮`	Ends equation.
[↱] [SHOW]	`CK=8F47` `LN=17`	Hold down SHOW to display checksum and length of equation. If numbers displayed match these, equation is correctly entered.

Equation 33-34: Stopping Sight Distance on Crest Vertical Curve

$$S < L: \quad L = \frac{AS^2}{200\left(\sqrt{h_1} + \sqrt{h_2}\right)^2}$$

$$S > L: \quad L = 2S - \frac{200\left(\sqrt{h_1} + \sqrt{h_2}\right)^2}{A}$$

Entered as

```
L=A×SQ(S)÷200÷SQ(SQRT(G)+SQRT(H))
L=2×S-200×SQ(SQRT(G)+SQRT(H))÷A
```

Discipline Transportation

Variables (See How to Use This Book for more information.)

$L = L$	length of curve	ft	m
$A = A$	absolute value of change in gradient	decimal	decimal
$S = S$	sight distance	ft	m
$G = h_1$	height of driver's eyes	3.5 ft	1.080 m
$H = h_2$	height of object sighted	2.0 ft	0.600 m

Keys	Display	Description
[↱] [EQN] [RCL] [L] [↱] [=]	`L=▮`	Selects Equation mode and starts new equation. *L* is length of curve.

[RCL] [A]	L=A■	*A* is change in gradient.
[×] [x^2]	L=A×SQ(■	Opens square function.
[RCL] [S]	L=A×SQ(S■	*S* is sight distance.
[↱] [)]	L=A×SQ(S)■	Closes square function.
[÷] [2] [0] [0]	=A×SQ(S)÷ 200_	Cursor changes for entering digits. Equation is now longer than the hp 33s display. For instructions on scrolling left and right, see How to Use This Book.
[÷] [x^2] [$\sqrt{x}$]	÷200÷SQ(SQRT(■	Opens square and square root functions. Functions can be nested.
[RCL] [G]	200÷SQ(SQRT(G■	*G* is height of driver's eyes (h_1).
[↱] [)] [+] [$\sqrt{x}$] [RCL] [H]	QRT(G)+SQRT(H■	*H* is height of object sighted (h_2).
[↱] [)] [↱] [)]	(G)+SQRT(H))■	Closes square function.
[ENTER]	L=A×SQ(S)÷200÷	Ends equation.
[↱] [SHOW]	CK=97D1 LN=33	Hold down SHOW to display checksum and length of equation. If numbers displayed match these, equation is correctly entered.

Keys	**Display**	**Description**
[↱] [EQN] [RCL] [L] [↱] [=]	L=■	Selects Equation mode and starts new equation. *L* is length of curve.
[2]	L= 2_	Cursor changes for entering digits.
[×] [RCL] [S]	L=2×S■	*S* is sight distance.

[−] [2] [0] [0]	L=2×S- 200_	
[×] [x^2]	L=2×S-200×SQ(▮	Opens square function.
[$\sqrt{x}$]	-200×SQ(SQRT(▮	Opens square root function. Equation is now longer than the hp 33s display. For instructions on scrolling left and right, see How to Use This Book.
[RCL] [G]	200×SQ(SQRT(G▮	*G* is height of driver's eyes (h_1).
[↱] [)] [+] [$\sqrt{x}$] [RCL] [H]	QRT(G)+SQRT(H▮	*H* is height of object sighted (h_2).
[↱] [)] [↱] [)]	T(G)+SQRT(H))▮	
[÷] [RCL] [A]	+SQRT(H))÷A▮	*A* is change in gradient.
[ENTER]	L=2×S-(200×SQ(	Ends equation.
[↱] [SHOW]	CK=7929 LN=31	Hold down SHOW to display checksum and length of equation. If numbers displayed match these, equation is correctly entered.

Equation 35-36: Stopping Sight Distance on Sag Vertical Curve

$$S < L: \quad L = \frac{AS^2}{X + 3.5S}$$

$$S > L: \quad L = 2S - \left(\frac{X + 3.5S}{A}\right)$$

Entered as

```
L=A×SQ(S)÷(X+3.5×S)
L=2×S-(X+3.5×S)÷A
```

Discipline Transportation

Variables (See How to Use This Book for more information.)

$L = L$	length of curve	ft	m
$A = A$	change in gradient	decimal	decimal
$S = S$	sight distance	ft	m
$X = X$	adjustment factor	200	120

Notes When solving, enter 200 for X when using feet, or 120 when using meters.

Keys	Display	Description
[↱] [EQN] [RCL] [L] [↱] [=]	L=▮	Selects Equation mode and starts new equation. *L* is length of curve.
[RCL] [A]	L=A▮	*A* is change in gradient.
[×] [x^2]	L=A×SQ(▮	Opens square function.
[RCL] [S]	L=A×SQ(S▮	*S* is sight distance.
[↱] [)]	L=A×SQ(S)▮	Closes square function.
[÷] [↱] [(] [RCL] [X]	L=A×SQ(S)÷(X▮	*X* is an adjustment factor, as explained above.
[+] [3] [·] [5]	×SQ(S)÷(X+ 3.5_	Cursor changes for entering digits. Equation is now longer than the hp 33s display. For instructions on scrolling left and right, see How to Use This Book.
[×] [RCL] [S] [↱] [)]	Q(S)÷(X+3.5×S)▮	
[ENTER]	L=A×SQ(S)÷(X+3.	Ends equation.
[↱] [SHOW]	CK=3FA9 LN=19	Hold down SHOW to display checksum and length of equation. If numbers displayed match these, equation is correctly entered.

Keys	Display	Description
[↱] [EQN] [RCL] [L] [↱] [=]	`L=▮`	Selects Equation mode and starts new equation. *L* is length of curve.
[2]	`L= 2_`	Cursor changes for entering digits.
[×] [RCL] [S]	`L=2×S▮`	*S* is sight distance.
[−] [↱] [(] [RCL] [X]	`L=2×S-(X▮`	*X* is an adjustment factor, as explained above.
[+] [3] [·] [5]	`L=2×S-(X+ 3.5▮`	
[×] [RCL] [S] [↱] [)]	`=2×S-(X+3.5×S)▮`	Equation is now longer than the hp 33s display. For instructions on scrolling left and right, see How to Use This Book.
[÷] [RCL] [A]	`×S-(X+3.5×S)÷A▮`	*A* is change in gradient.
[ENTER]	`L=2×S-(X+3.5×S)`	Ends equation.
[↱] [SHOW]	`CK=7EE3` `LN=17`	Hold down SHOW to display checksum and length of equation. If numbers displayed match these, equation is correctly entered.

Equation 37: Heavy Vehicle Adjustment Factor

$$f_{\mathrm{HV}} = \frac{1}{1 + P_T\,(E_T - 1) + P_R\,(E_R - 1)}$$

Entered as

`F=1÷(1+P×(T-1)+R×(E-1))`

Discipline Transportation

Variables (See How to Use This Book for more information.)

$F = f_{\mathrm{HV}}$	heavy vehicle adjustment factor	–
$P = P_T$	percentage of traffic, trucks and buses	%

$R = P_R$	percentage of traffic, RVs	%
$T = E_T$	passenger car equivalent, trucks and buses	–
$E = E_R$	passenger car equivalent, RVs	–

Keys	Display	Description
[↱] [EQN] [RCL] [F] [↱] [=]	`F=▮`	Selects Equation mode and starts new equation. *F* is heavy vehicle adjustment factor (f_{HV}).
[1]	`F= 1_`	Cursor changes for entering digits.
[÷] [↱] [(] [1] [+] [RCL] [P]	`F=1÷(1+P▮`	*P* is percentage of traffic that is trucks and buses (P_T).
[×] [↱] [(] [RCL] [T]	`F=1÷(1+P×(T▮`	*T* is passenger car equivalent for trucks and buses (E_T).
[−] [1]	`=1÷(1+P×(T- 1_`	Equation is now longer than the hp 33s display. For instructions on scrolling left and right, see How to Use This Book.
[↱] [)] [+] [RCL] [R]	`÷(1+P×(T-1)+R▮`	*R* is percentage of traffic that is RVs (P_R).
[×] [↱] [(] [RCL] [E]	`+P×(T-1)+R×(E▮`	*E* is passenger car equivalent for RVs (E_R).
[−] [1] [↱] [)] [↱] [)]	`T-1)+R×(E-1))▮`	
[ENTER]	`F=1÷(1+P×(T-1)`	Ends equation.
[↱] [SHOW]	`CK=52D9` `LN=23`	Hold down SHOW to display checksum and length of equation. If numbers displayed match these, equation is correctly entered.

Equation 38: Constant Acceleration

$$s = s_0 + v_0 t + \frac{a_0 t^2}{2}$$

Entered as

```
S=D+V×T+A×SQ(T)÷2
```

Discipline Transportation

Variables (See How to Use This Book for more information.)

$S = s$	distance	ft	m
$D = s_0$	initial distance	ft	m
$V = v_0$	initial velocity	ft/sec	m/s
$T = t$	time	sec	s
$A = a_0$	acceleration	ft/sec^2	m/s^2

Keys	Display	Description
[↱] [EQN] [RCL] [S] [↱] [=]	S=■	Selects Equation mode and starts new equation. *S* is distance traveled (*s*).
[RCL] [D]	S=D■	*D* is initial distance (s_0).
[+] [RCL] [V]	S=D+V■	*V* is initial velocity (v_0).
[×] [RCL] [T]	S=D+V×T■	*T* is time (*t*).
[+] [RCL] [A]	S=D+V×T+A■	*A* is acceleration (*a*).
[×] [x^2]	S=D+V×T+A×SQ(■	Opens square function.
[RCL] [T] [↱] [)]	D+V×T+A×SQ(T)■	Enters *T* and closes square function. Equation is now longer than the hp 33s display. For instructions on scrolling left and right, see How to Use This Book.
[÷] [2]	×T+A×SQ(T) 2_	Cursor changes for entering digits.
[ENTER]	S=D+V×T+A×SQ(T	Ends equation.

[⮫] [SHOW]	CK=FED0 LN=17	Hold down SHOW to display checksum and length of equation. If numbers displayed match these, equation is correctly entered.

Equation 39: Average Control Delay

$$d = 0.50C\left(\frac{\left(1-\frac{g}{C}\right)^2}{1-\left(\min(1,X)\left(\frac{g}{C}\right)\right)}\right)\left(\frac{(1-P)\,f_{\text{PA}}}{1-\frac{g}{C}}\right) + 900T\left((X-1)+\sqrt{(X-1)^2+\left(\frac{8kIX}{cT}\right)}\right)$$

Entered as

```
D=0.5×C×SQ(1-G÷C)÷(1-M×G÷C)×(1-P)×F÷(1-G÷C)
     +900×T×(X-1+SQRT(SQ(X-1)+8×K×I×X÷Z÷T))
```

Discipline Transportation

Variables (See How to Use This Book for more information.)

$D = d$	average control delay	sec	s
$C = C$	cycle length	sec	s
$G = g$	effective green time	sec	s
$M = \min(1, X)$	lesser of 1 and X	–	–
$P = P$	proportion of vehicles arriving on green	–	–
$F = f_{\text{PA}}$	supplemental adjustment factor for when platoon arrives during green	–	–
$T = T$	duration of analysis period	hr	h
$X = X$	volume-capacity ratio of lane group	–	–
$K = k$	incremental delay factor, dependent on controller settings	–	–
$I = I$	upstream filtering/metering adjustment factor	–	–
$Z = c$	lane group capacity	vph	vph

Notes This is an expansion of the formula $d = d_1\,(\text{PF}) + d_2 + d_3$, assuming no initial queue (making d_3 equal to 0 s.)

Keys	Display	Description
[⮫] [EQN] [RCL] [D] [⮫] [=]	D=▮	Selects Equation mode and starts new equation. *D* is average control delay (*d*).

Keys	Display	Description
[.] [5]	`D= 0.5_`	Cursor changes for entering digits.
[×] [RCL] [C]	`D=0.5×C■`	*C* is cycle length.
[×] [x^2]	`D=0.5×C×SQ(■`	Opens square function.
[1] [−] [RCL] [G]	`D=0.5×C×SQ(1-G■`	*G* is effective green time (*g*).
[÷] [RCL] [C]	`0.5×C×SQ(1-G÷C■`	Equation is now longer than the hp 33s display. For instructions on scrolling left and right, see How to Use This Book.
[↱] [)] [÷] [↱] [(] [1] [−] [RCL] [M]	`Q(1-G÷C)÷(1-M■`	*M* is the lesser of 1 and *X*.
[×] [RCL] [G] [÷] [RCL] [C] [↱] [)]	`÷C)÷(1-M×G÷C)■`	
[×] [↱] [(] [1] [−] [RCL] [P]	`1-M×G÷C)×(1-P■`	*P* is proportion of vehicles arriving on green.
[↱] [)] [×] [RCL] [F]	`×G÷C)×(1-P)×F■`	*F* is supplemental adjustment factor for when platoon arrives during green (f_{PA}).
[÷] [↱] [(] [1] [−] [RCL] [G] [÷] [RCL] [C] [↱] [)]	`-P)×F÷(1-G÷C)■`	
[+] [9] [0] [0] [×] [RCL] [T]	`(1-G÷C)+900×T■`	*T* is duration of analysis period.
[×] [↱] [(] [RCL] [X]	`G÷C)+900×T×(X■`	*X* is volume-capacity ratio of lane group.

[−] [1] [+] [√x]	×T×(X-1+SQRT(■	Opens square root function.
[x²]	(X-1+SQRT(SQ(■	Opens square function. Functions can be nested.
[RCL] [X] [−] [1] [↱] [)] [+] [8] [×] [RCL] [K]	T(SQ(X-1)+8×K■	*K* is incremental delay factor (*k*).
[×] [RCL] [I]	SQ(X-1)+8×K×I■	*I* is upstream filtering/metering adjustment factor.
[×] [RCL] [X] [÷] [RCL] [Z]	Z -1)+8×K×I×X÷Z■	*Z* is lane group capacity (*c*).
[÷] [RCL] [T] [↱] [)] [↱] [)]	8×K×I×X÷Z÷T))■	
[ENTER]	D=0.5×C×SQ(1-G÷	Ends equation.
[↱] [SHOW]	CK=CDB8 LN=81	Hold down SHOW to display checksum and length of equation. If numbers displayed match these, equation is correctly entered.

Equation 40: Free Flow Speed

$$\text{FFS} = S_{\text{FM}} + 0.00776\left(\frac{V_f}{f_{\text{HV}}}\right)$$

Entered as

`F=S+0.00776×V÷H`

Discipline Transportation

Variables (See How to Use This Book for more information.)

$F = \text{FFS}$	free flow speed	mph	km/h
$S = S_{\text{FM}}$	mean traffic speed measured in the field	mph	km/h
$V = V_f$	observed flow rate	vph	vph
$H = f_{\text{HV}}$	heavy vehicle adjustment factor	–	–

Keys	Display	Description
[↱] [EQN] [RCL] [F] [↱] [=]	`F=■`	Selects Equation mode and starts new equation. *F* is free flow speed (FFS).
[RCL] [S]	`F=S■`	*S* is mean traffic speed (S_{FM}).
[+] [.] [0] [0] [7] [7] [6]	`F=S+ 0.00776_`	Cursor changes for entering digits.
[×] [RCL] [V]	`F=S+0.00776×V■`	*V* is observed flow rate (V_f).
[÷] [RCL] [H]	`=S+0.00776×V÷H■`	*H* is heavy vehicle adjustment factor (f_{HV}). Equation is now longer than the hp 33s display. For instructions on scrolling left and right, see How to Use This Book.
[ENTER]	`F=S+0.00776×V÷H`	Ends equation.
[↱] [SHOW]	`CK=2FF7` `LN=15`	Hold down SHOW to display checksum and length of equation. If numbers displayed match these, equation is correctly entered.

WATER RESOURCES

Equation 41: Manning Equation

$$Q = \frac{K}{n} A R^{2/3} S^{1/2}$$

Entered as

`Q=K÷N×A×R^(2÷3)×S^0.5`

Discipline Water Resources

Variables (See How to Use This Book for more information.)

$Q = Q$	quantity of flow	ft^3/sec	m^3/s
$K = K$	constant	1.49	1.00

$N = n$	Manning roughness coefficient	–	–
$A = A$	cross-sectional area	ft^2	m^2
$R = R$	hydraulic radius	ft	m
$S = S$	slope of energy line	decimal	decimal

Notes When solving, enter 1.49 for K if using U.S. customary units, and 1 if using SI units.

Keys	Display	Description
[↱] [EQN] [RCL] [Q] [↱] [=]	Q=▮	Selects Equation mode and starts new equation. *Q* is quantity of flow.
[RCL] [K]	Q=K▮	*K* is a constant (see Notes above).
[÷] [RCL] [N]	Q=K÷N▮	*N* is the Manning roughness coefficient (*n*).
[×] [RCL] [A]	Q=K÷N×A▮	*A* is cross-sectional area.
[×] [RCL] [R]	Q=K÷N×A×R▮	*R* is hydraulic radius.
[y^x]	Q=K÷N×A×R^▮	Enters an exponent.
[↱] [(] [2]	Q=K÷N×A×R^(2_	Cursor changes for entering digits.
[÷] [3]	K÷N×A×R^(2÷ 3_	Equation is now longer than the hp 33s display. For instructions on scrolling left and right, see How to Use This Book.
[↱] [)] [×] [RCL] [S]	N×A×R^(2÷3)×S▮	*S* is slope of energy line.
[y^x] [·] [5]	R^(2÷3)×S^ 0.5▮	0.50 is used for 1/2.
[ENTER]	Q=K÷N×A×R^(2÷3	Ends equation.
[↱] [SHOW]	CK=4A8B LN=21	Hold down SHOW to display checksum and length of equation. If numbers displayed match these, equation is correctly entered.

Equation 42: Energy Equation (Bernoulli Equation)

$$\frac{p_2}{\gamma} + z_2 + \frac{v_2^2}{2g} + h_f = \frac{p_1}{\gamma} + z_1 + \frac{v_1^2}{2g}$$

Entered as

```
P÷W+Z+SQ(V)÷2÷G+H=O÷W+Y+SQ(U)÷2÷G
```

Discipline Water Resources

Variables (See How to Use This Book for more information.)

$P = p_2$	pressure at point 2	lbf/ft^2	Pa
$W = \gamma$	specific weight	lbf/ft^3	N/m^3
$Z = z_2$	elevation at point 2	ft	m
$V = v_2$	velocity at point 2	ft/sec	m/s
$G = g$	gravitational acceleration	32.2 ft/sec^2	9.81 m/s^2
$H = h_f$	head loss due to friction	ft	m
$O = p_1$	pressure at point 1	lbf/ft^2	Pa
$Y = z_1$	elevation at point 1	ft	m
$U = v_1$	velocity at point 1	ft/sec	m/s

Notes This equation can be used to solve problems with or without head loss. If there are no friction losses to take into account, enter zero as the value for H. For problems in real fluid flow or steady, incompressible flow in conduits and pipes, H stands for h_f.

Keys	Display	Description
[↱] [EQN] [RCL] [P]	`P■`	Selects Equation mode and starts new equation. *P* is pressure (p_2).
[÷] [RCL] [W]	`P÷W■`	*W* is specific weight (γ).
[+] [RCL] [Z]	`P÷W+Z■`	*Z* is elevation (z_2).
[+] [x^2]	`P÷W+Z+SQ(■`	Opens square function.
[RCL] [V]	`P÷W+Z+SQ(V■`	*V* is velocity (v_2).
[↱] [)]	`P÷W+Z+SQ(V)■`	Closes square function.
[÷] [2]	`÷W+Z+SQ(V)÷ 2_`	Cursor changes for entering digits. Equation is now longer than the hp 33s display. For instructions on scrolling left and right, see How to Use This Book.

Keys	Display	Description
[÷] [RCL] [G]	`W+Z+SQ(V)÷2÷G█`	*G* is gravitational acceleration. When solving, enter 32.2 for U.S. units or 9.81 for SI units.
[+] [RCL] [H]	`Z+SQ(V)÷2÷G+H█`	*H* is head loss (h_f).
[↱] [=] [RCL] [O]	`SQ(V)÷2÷G+H=O█`	*O* stands for p_1, as explained above.
[÷] [RCL] [W] [+] [RCL] [Y]	`)÷2÷G+H=O÷W+Y█`	*Y* stands for z_1.
[+] [x^2] [RCL] [U]	`+H=O÷W+Y+SQ(U█`	*U* stands for v_1
[↱] [)] [÷] [2] [÷] [RCL] [G]	`W+Y+SQ(U)÷2÷G█`	
[ENTER]	`P÷W+Z+SQ(V)÷2÷`	Ends equation.
[↱] [SHOW]	`CK=5017` `LN=33`	Hold down SHOW to display checksum and length of equation. If numbers displayed match these, equation is correctly entered.

Equation 43: Darcy-Weisbach Equation

$$h_f = f\frac{Lv^2}{d2g}$$

Entered as

`H=F×L×SQ(V)÷D÷2÷G`

Discipline Water Resources

Variables (See How to Use This Book for more information.)

$H = h_f$	head loss due to friction	ft	m
$F = f$	Darcy friction factor	–	–
$L = L$	length	ft	m
$V = v$	velocity	ft/sec	m/s
$D = d$	diameter	ft	m
$G = g$	gravitational acceleration	32.2 ft/sec^2	9.81 m/s^2

Keys	Display	Description
[⮫] [EQN] [RCL] [H] [⮫] [=]	`H=■`	Selects Equation mode and starts new equation. *H* is head loss (h_f).
[RCL] [F]	`H=F■`	*F* is friction factor (*f*).
[×] [RCL] [L]	`H=F×L■`	*L* is length.
[×] [x^2]	`H=F×L×SQ(■`	Opens square function.
[RCL] [V]	`H=F×L×SQ(V■`	*V* is velocity (v).
[⮫] [)]	`H=F×L×SQ(V)■`	Closes square function.
[÷] [RCL] [D]	`H=F×L×SQ(V)÷D■`	*D* is pipe diameter.
[÷] [2]	`×L×SQ(V)÷D÷ 2_`	Cursor changes for entering digits. Equation is now longer than the hp 33s display. For instructions on scrolling left and right, see How to Use This Book.
[÷] [RCL] [G]	`L×SQ(V)÷D÷2÷G■`	*G* is gravitational acceleration (*g*). When solving, enter 32.2 with U.S. units or 9.81 with SI units.
[ENTER]	`H=F×L×SQ(V)÷D÷`	Ends equation.
[⮫] [SHOW]	`CK=7FAB` `LN=17`	Hold down SHOW to display checksum and length of equation. If numbers displayed match these, equation is correctly entered.

Equation 44: Specific Speed of Pump

$$n_s = \frac{n\sqrt{Q}}{h_A^{0.75}}$$

Entered as

`S=N×SQRT(Q)÷H^0.75`

Discipline Water Resources

Variables (See How to Use This Book for more information.)

$S = n_s$	specific rotational speed	–
$N = n$	rotational speed	rpm
$Q = Q$	flow rate	gpm
$H = h_A$	head added by pump	ft

Keys	Display	Description
[⮫] [EQN] [RCL] [S] [⮫] [=]	`S=▮`	Selects Equation mode and starts new equation. *S* is specific speed (n_s).
[RCL] [N]	`S=N▮`	*N* is rotational speed (*n*).
[×] [√x]	`S=N×SQRT(▮`	Opens square root function.
[RCL] [Q]	`S=N×SQRT(Q▮`	*Q* is flow rate.
[⮫] [)]	`S=N×SQRT(Q)▮`	Closes square root function.
[÷] [RCL] [H]	`S=N×SQRT(Q)÷H▮`	*H* is head added by pump (h_A).
[yˣ] [·] [7] [5]	`QRT(Q)÷H^0.75_`	Cursor changes for entering digits. Equation is now longer than the hp 33s display. For instructions on scrolling left and right, see How to Use This Book.
[ENTER]	`S=N×SQRT(Q)÷H^`	Ends equation.
[⮫] [SHOW]	`CK=2DA6` `LN=18`	Hold down SHOW to display checksum and length of equation. If numbers displayed match these, equation is correctly entered.

Equation 45: Hydraulic Radius of Trapezoidal Channel

$$R = \frac{bd\sin\theta + d^2\cos\theta}{b\sin\theta + 2d}$$

Entered as

`R=(B×D×SIN(T)+SQ(D)×COS(T))÷(B×SIN(T)+2×D)`

Discipline Water Resources

Variables (See How to Use This Book for more information.)

$R = R$	hydraulic radius	ft	m
$B = b$	channel width at base	ft	m
$D = d$	channel depth	ft	m
$T = \theta$	angle of channel side from horizon	deg	deg

Keys	Display	Description
[↱] [EQN] [RCL] [R] [↱] [=]	`R=▮`	Selects Equation mode and starts new equation. *R* is hydraulic radius.
[↱] [(] [RCL] [B]	`R=(B▮`	*B* is channel width at base (*b*).
[×] [RCL] [D]	`R=(B×D▮`	*D* is channel depth (*d*).
[×] [SIN]	`R=(B×D×SIN(▮`	Opens sine function.
[RCL] [T]	`R=(B×D×SIN(T▮`	*T* is angle of channel side from horizon (θ).
[↱] [)]	`R=(B×D×SIN(T)▮`	Closes sine function.
[+] [x^2]	`×D×SIN(T)+SQ(▮`	Opens square function. Equation is now longer than the hp 33s display. For instructions on scrolling left and right, see How to Use This Book.
[RCL] [D] [↱] [)]	`×SIN(T)+SQ(D)▮`	Enters *D* and closes square function.
[×] [COS]	`T)+SQ(D)×COS(▮`	Opens cosine function.
[RCL] [T] [↱] [)]	`+SQ(D)×COS(T)▮`	Enters *T* and closes cosine function.

Keys	Display	Description
[↱] [)] [÷]	D)×COS(T))÷(B■	
[↱] [(] [RCL] [B]		
[×] [SIN] [RCL] [T]	(T))÷(B×SIN(T■	
[↱] [)] [+] [2]	÷(B×SIN(T)+ 2_	Cursor changes for entering digits.
[×] [RCL] [D] [↱] [)]	(B×SIN(T)+2×D)■	
[ENTER]	R=(B×D×SIN(T)+×	Ends equation.
[↱] [SHOW]	CK=AA0D LN=42	Hold down SHOW to display checksum and length of equation. If numbers displayed match these, equation is correctly entered.

Equations 46-47: Confined Stream in Pipe Bend

$$F_x = p_2 A_2 \cos\theta - p_1 A_1 - \frac{\dot{m}\left(\mathrm{v}_2 \cos\theta - \mathrm{v}_1\right)}{g_c}$$

$$F_y = \left(p_2 A_2 + \frac{\dot{m}\mathrm{v}_2}{g_c}\right)\sin\theta$$

Entered as

```
X=P×A×COS(T)-O×Z+M×(V×COS(T)-U)÷G
Y=(P×A+M×V÷G)×SIN(T)
```

Discipline Water Resources

Variables (See How to Use This Book for more information.)

$X = F_x$	x-component of force	lbf	N
$O = p_1$	pressure at point 1	lbf/ft^2	Pa
$P = p_2$	pressure at point 2	lbf/ft^2	Pa
$Z = A_1$	cross-sectional area at point 1	ft^2	m^2
$A = A_2$	cross-sectional area at point 2	ft^2	m^2
$M = \dot{m}$	mass flow rate	lbm/sec	kg/s
$Y = F_y$	y-component of force	lbf	N
$U = \mathrm{v}_1$	velocity at point 1	ft/sec	m/s

$G = g_c$	gravitational constant	32.2 lbm-ft/lbf-sec^2	1
$V = v_2$	velocity at point 2	ft/sec	m/s
$T = \theta$	pipe bend angle	deg	deg

Notes These equations can be used with either U.S. or SI units. With U.S. units, enter 32.2 for G. With SI units, enter 1.

Keys	Display	Description
[⮫] [EQN] [RCL] [X] [⮫] [=]	X=▮	Selects Equation mode and starts new equation. *X* is *x*-component of force (F_x).
[RCL] [P]	X=P▮	*P* is pressure at point 2 (p_2).
[×] [RCL] [A]	X=P×A▮	*A* is area at point 2 (A_2).
[×] [COS]	X=P×A×COS(▮	Opens cosine function.
[RCL] [T]	X=P×A×COS(T▮	*T* is pipe bend angle (θ).
[⮫] [)]	X=P×A×COS(T)▮	Closes cosine function.
[−] [RCL] [O]	=P×A×COS(T)-O▮	*O* is pressure at point 1 (p_1). Equation is now longer than the hp 33s display. For instructions on scrolling left and right, see How to Use This Book.
[×] [RCL] [Z]	×A×COS(T)-O×Z▮	*Z* is area at point 1 (A_1).
[+] [RCL] [M]	×COS(T)-O×Z+M▮	*M* is mass flow rate ($\dot{m}$).
[×] [⮫] [(] [RCL] [V]	S(T)-O×Z+M×(V▮	*V* is velocity at point 2 (v_2).
[×] [COS] [RCL] [T] [⮫] [)] [−] [RCL] [U]	M×(V×COS(T)-U▮	*U* is velocity at point 1 (v_1).
[⮫] [)] [÷] [RCL] [G]	OS(T)-U)÷G	*G* is gravitational constant. (See Notes above.)
[ENTER]	X=P×A×COS(T)-O	Ends equation.

[⮫] [SHOW]	CK=F512 LN=33	Hold down SHOW to display checksum and length of equation. If numbers displayed match these, equation is correctly entered.

Keys	Display	Description
[⮫] [EQN] [RCL] [Y] [⮫] [=]	Y=■	Selects Equation mode and starts new equation. *Y* is *y*-component of force (F_y).
[⮫] [(] [RCL] [P]	Y=(P■	*P* is pressure at point 2 (p_2).
[×] [RCL] [A]	Y=(P×A■	*A* is area at point 2 (A_2).
[+] [RCL] [M]	Y=(P×A+M■	*M* is mass flow rate ($\dot{m}$).
[×] [RCL] [V]	Y=(P×A+M×V■	*V* is velocity at point 2 (v_2).
[÷] [RCL] [G]	Y=(P×A+M×V÷G■	*G* is gravitational constant. (See Notes above.)
[⮫] [)] [×] [SIN]	A+M×V÷G)×SIN(■	Opens sine function. Equation is now longer than the hp 33s display. For instructions on scrolling left and right, see How to Use This Book.
[RCL] [T] [⮫] [)]	M×V÷G)×SIN(T)■	Enters *T* and closes sine function. *T* is pipe bend angle (θ).
[ENTER]	Y=(P×A+M×V÷G)×	Ends equation.
[⮫] [SHOW]	CK=1F28 LN=20	Hold down SHOW to display checksum and length of equation. If numbers displayed match these, equation is correctly entered.

Equation 48: Conjugate Depths for Rectangular Channel

$$d_1 = -\tfrac{1}{2}d_2 + \sqrt{\frac{2v_2^2 d_2}{g} + \frac{d_2^2}{4}}$$

$$d_2 = -\tfrac{1}{2}d_1 + \sqrt{\frac{2v_1^2 d_1}{g} + \frac{d_1^2}{4}}$$

Entered as

```
C=-0.5×D+SQRT(2×SQ(V)×D÷G+SQ(D)÷4)
```

Discipline Water Resources

Variables (See How to Use This Book for more information.)

$C = d_1$	depth at point 1	ft	m
$D = d_2$	depth at point 2	ft	m
$V = v_2$	velocity at point 2	ft/sec	m/s
$G = g$	gravitational acceleration	32.2 ft/sec^2	9.81 m/s^2

Notes The equations for d_1 and d_2 are symmetrical. The equation as entered is based on that for d_1, but can be used for d_2 by exchanging points 1 and 2.

Keys	Display	Description
[↱] [EQN] [RCL] [C] [↱] [=]	`C=▮`	Selects Equation mode and starts new equation. *C* is depth at point 1 (d_1).
[+/−]	`C=-`	Changes sign of next number.
[.] [5]	`C=- 0.5_`	Cursor changes for entering digits.
[×] [RCL] [D]	`C=-0.5×D▮`	*D* is depth at point 2 (d_2).
[+] [√x]	`C=-0.5×D+SQRT(▮`	Opens square root function.
[2] [×] [x²]	`×D+SQRT(2×SQ(▮`	Opens square function. Equation is now longer than the hp 33s display. For instructions on scrolling left and right, see How to Use This Book.
[RCL] [V]	`D+SQRT(2×SQ(V▮`	*V* is velocity at point 2 (v_2).
[↱] [)]	`+SQRT(2×SQ(V)▮`	Closes square function.

Keys	Display	Description
[×] [RCL] [D] [÷] [RCL] [G]	T(2×SQ(V)×D÷G▮	*G* is gravitational acceleration (*g*). When solving, enter 32.2 with U.S. units or 9.81 with SI units.
[+] [x^2]	SQ(V)×D÷G+SQ(▮	
[RCL] [D] [↱] [)] [÷] [4] [↱] [)]	×D÷G+SQ(D)÷4)▮	Closes square root function.
[ENTER]	C=-0.5×D+SQRT(2	Ends equation.
[↱] [SHOW]	CK=CAD1 LN=34	Hold down SHOW to display checksum and length of equation. If numbers displayed match these, equation is correctly entered.

Equation 49: FAA Formula for Time to Concentration

$$t_c = \frac{(1.8)\,(1.1 - C)\,\sqrt{L_o}}{S^{1/3}}$$

Entered as

`T=1.8×(1.1-C)×SQRT(L)÷S^(1÷3)`

Discipline Water Resources

Variables (See How to Use This Book for more information.)

$T = t_c$	time to concentration	min
$C = C$	rational runoff coefficient	–
$S = S$	slope	%
$L = L_o$	length	ft

Notes When solving, enter slope as a percentage.

Keys	Display	Description
[↱] [EQN] [RCL] [T] [↱] [=]	T=▮	Selects Equation mode and starts new equation. *T* is time to concentration (t_c).
[1] [.] [8]	T= 1.8_	Cursor changes for entering digits.

[×] [↱] [(] [1] [·] [1]	`T=1.8×( 1.1_`	
[−] [RCL] [C]	`T=1.8×(1.1-C▮`	*C* is rational runoff coefficient.
[↱] [)] [×] [√x]	`×(1.1-C)×SQRT(▮`	Opens square root function. Equation is now longer than the hp 33s display. For instructions on scrolling left and right, see How to Use This Book.
[RCL] [L]	`(1.1-C)×SQRT(L▮`	*L* is overland length (L_o).
[↱] [)]	`1.1-C)×SQRT(L)▮`	Closes square root function.
[÷] [RCL] [S]	`-C)×SQRT(L)÷S▮`	*S* is slope.
[y^x]	`C)×SQRT(L)÷S^▮`	Enters an exponent.
[↱] [(] [1] [÷] [3] [↱] [)]	`RT(L)÷S^(1÷3)▮`	
[ENTER]	`C=−0.5×D+SQRT(2`	Ends equation.
[↱] [SHOW]	`CK=4E11` `LN=29`	Hold down SHOW to display checksum and length of equation. If numbers displayed match these, equation is correctly entered.

Equation 50: Throat Velocity in Venturi Meter

$$v_2 = \left(\frac{C_v}{\sqrt{1-\beta^4}}\right)\sqrt{\frac{2g\left(\rho_m-\rho\right)h}{\rho}}$$

Entered as

`V=C÷SQRT(1-B^4)×SQRT(2×G×(M-P)×H÷P)`

Discipline Water Resources

Variables (See How to Use This Book for more information.)

$V = v_2$	throat velocity	ft/sec	m/s
$C = C_v$	velocity coefficient	–	–

$B = \beta$	beta ratio (throat diameter/upstream diameter)	–	–
$G = g$	gravitational acceleration	32.2 ft/sec^2	9.81 m/s^2
$M = \rho_m$	manometer fluid density	lbm/ft^3	kg/m^3
$P = \rho$	water density	lbm/ft^3	kg/m^3
$H = h$	manometer reading	ft	m

Keys	Display	Description
[↱] [EQN] [RCL] [V] [↱] [=]	V=▮	Selects Equation mode and starts new equation. *V* is throat velocity (v_2).
[RCL] [C]	V=C▮	*C* is velocity coefficient (C_v).
[÷] [√x]	V=C÷SQRT(▮	Opens square root function.
[1]	V=C÷SQRT(1_	Cursor changes for entering digits.
[−] [RCL] [B]	V=C÷SQRT(1-B▮	B is beta ratio (β).
[y^x]	V=C÷SQRT(1-B^▮	Enters an exponent.
[4]	C÷SQRT(1-B^ 4_	B^4 is entered as B^4. Equation is now longer than the hp 33s display. For instructions on scrolling left and right, see How to Use This Book.
[↱] [)] [×] [√x] [2] [×] [RCL] [G]	B^4)×SQRT(2×G▮	*G* is gravitational acceleration (*g*). When solving, enter 32.2 with U.S. units or 9.81 with SI units.
[×] [↱] [(] [RCL] [M]	)×SQRT(2×G×(M▮	*M* is manometer fluid density (ρ_m).
[−] [RCL] [P]	SQRT(2×G×(M-P▮	*P* is water density (ρ).
[↱] [)] [×] [RCL] [H]	T(2×G×(M-P)×H▮	*H* is manometer reading (*h*).
[÷] [RCL] [P] [↱] [)]	×G×(M-P)×H÷P)▮	
[ENTER]	V=C÷SQRT(1-B^4	Ends equation.

[↱] [SHOW]	CK=A3E3 LN=35	Hold down SHOW to display checksum and length of equation. If numbers displayed match these, equation is correctly entered.

PRACTICE PROBLEMS USING STORED EQUATIONS

B

Part A presented instructions for storing 50 equations in the hp33s calculator. The 15 practice problems in Part B show how to save time when solving problems by using stored equations. These problems are taken from a variety of sources (a list is given in the reference section of this book) and are typical of the problems you can expect on the civil PE exam.

To use a stored equation, first enter Equation mode. Scroll through the list of equations to bring the one you want into the bottom row of the display.

If the equation has only a single variable to the left of the equal sign, and you want to solve for that variable, press ENTER. To solve for any other variable, press the SOLVE key (one of the silver keys just under the display) followed by the letter name of the variable.

Once you've done that, the hp33s will prompt you for the other variables. At each prompt, enter the value for that variable, followed by the [R/S] key. You can include operators, so pressing [5] [ENTER] [4] [y^x] [R/S] will enter the value 5^4 or 625. This can be convenient, as several of the problems in this part show.

In Part A, only the bottom line of the display is shown in the instructions (except for showing checksums). Here in Part B, both the top and bottom lines are shown whenever the top line is a prompt for the desired variable and the bottom line is the current value of that variable. When the current value is the one you want, you can retain it without needing to enter it again. This can save time when solving a series of equations involving the same values.

The displays shown in these instructions are based on the assumption that all variables equal zero at the start of solving each problem. But in actual use—unless you were to clear the calculator before every problem—you will see whatever value you most recently assigned to that variable. This doesn't affect the steps in the solution.

PROBLEM 1

An industrial plant proposes to return noncontact cooling water to a river. The river water temperature upstream of the mixing zone is 6°C and the returned cooling water temperature is 20°C. The river flows at 280 m^3/s and the cooling water flow is 11 m^3/s. The reaeration and deoxygenation river constants are equal at 0.080 d^{-1}. What are most nearly the river constants downstream of the mixing zone?

(A) 0.075 d^{-1}
(B) 0.085 d^{-1}
(C) 0.19 d^{-1}
(D) 0.47 d^{-1}

SOLUTION 1

The equation for mass balance (Equation 1, p. 9) can be used to determine the temperature of the mixed flows.

Keys	Display	Description
[↱] [EQN]	F=(B×P+C×Q)÷(P	Enters Equation mode. Scroll through list to bring needed equation into bottom line of display.
[ENTER]	B? 0.0000	Starts to solve for *F*. Prompts for *B* and displays current value.
[6] [R/S]	P? 0.0000	Enters 6 for *B*. Prompts for *P* and displays current value.
[2] [8] [0] [R/S]	C? 0.0000	Enters 280 for *P*. Prompts for *C* and displays current value.
[2] [0] [R/S]	Q? 0.0000	Enters 20 for *C*. Prompts for *Q* and displays current value.
[1] [1] [R/S]	F = 6.5292	Enters 11 for *Q* and solves for *F*.

The temperature of the mixed flow is 6.5°C.

The equation for temperature correction (Equation 7, p. 17) can be used to find the river constants. For temperatures between 4°C and 20°C, use 1.135 for the temperature constant.

Keys	Display	Description
[⮫] [EQN]	J=K×R^(S-T)	Enters Equation mode. Scroll through list to bring needed equation into bottom line of display.
[ENTER]	K? 0.0000	Starts to solve for *J*. Prompts for *K* and displays current value.
[.] [0] [8] [R/S]	R? 0.0000	Enters 0.080 for *K*. Prompts for *R* and displays current value.
[1] [.] [1] [3] [5] [R/S]	S? 0.0000	Enters 1.135 for *R*. Prompts for *S* and displays current value.
[6] [.] [5] [R/S]	T? 0.0000	Enters 6.5 for *S*. Prompts for *T* and displays current value.
[6] [R/S]	J = 0.0852	Enters 6 for *T* and solves for *J*.

The answer is (B).

PROBLEM 2

A stream segment below a stormwater discharge exhibits the following characteristics.

saturated dissolved oxygen concentration	10.9 mg/L
mixed ultimate biochemical oxygen demand (BOD_u) at the discharge	7.2 mg/L
dissolved oxygen deficit at the discharge point	3.2 mg/L
reaeration constant	0.07 d^{-1}
deoxygenation constant	0.04 d^{-1}

What is most nearly the time to the critical point of the oxygen sag curve?

(A) 5.1 d
(B) 8.3 d
(C) 10 d
(D) 14 d

SOLUTION 2

Use the formula for the time to the critical point (Equation 5, p. 14).

Keys	Display	Description
[↱] [EQN]	T=LOG((D×U−R×O	Enters Equation mode. Scroll through list to bring needed equation to bottom line of display.
[ENTER]	D? 0.0000	Starts to solve for *T*. Prompts for *D* and displays current value.
[·] [0] [1] [7] [R/S]	U? 0.0000	Enters 0.017 for *D*. Prompts for *U* and displays current value.
[7] [·] [2] [R/S]	R? 0.0000	Enters 7.2 for *U*. Prompts for *R* and displays current value.
[·] [0] [3] [R/S]	O? 0.0000	Enters 0.03 for *R*. Prompts for *O* and displays current value.
[3] [·] [2] [R/S]	T = 5.1002	Enters 3.2 for *O* and solves for *T*.

The answer is (A).

PROBLEM 3

What is most nearly the required width of the rectangular horizontal-flow grit chamber where the following conditions apply?

flow rate	3.5×10^6 gal/day
depth	4 ft
mean particle diameter	0.22 mm
grit specific gravity	2.65
Camp constant	0.05
Darcy friction factor	0.03

(A) 0.60 ft
(B) 1.9 ft
(C) 8.8 ft
(D) 14 ft

SOLUTION 3

Use the formula for velocity of settling (Equation 6, p. 16) to find the horizontal velocity. The answer is to be in feet, so take care to use consistent units. Convert D to feet and enter 32.2 as the value of G. The conversion to feet can be calculated during the solving process, as shown below. It could also be calculated separately before solving.

Keys	Display	Description
[↱] [EQN]	V=SQRT(8×K×G×D	Enters Equation mode. Scroll through list to bring needed equation to bottom line of display.
[ENTER]	K? 0.0000	Tells calculator to solve for *V*. Prompts for *K* and displays current value.
[.] [0] [5] [R/S]	G? 0.0000	Enters 0.05 for *K*. Prompts for *G* and displays current value.
[3] [2] [.] [2] [R/S]	D? 0.0000	Enters 32.2 for *G*. Prompts for *D* and displays current value.
[.] [2] [2] [ENTER] [.] [0] [0] [3] [2] [8] [×] [R/S]	F? 0.0000	Converts 0.22 mm to feet and enters the result for *D*. Prompts for *F* and displays current value.
[.] [0] [3] [R/S]	S? 0.0000	Enters 0.03 for *F*. Prompts for *S* and displays current value.
[2] [.] [6] [5] [R/S]	V = 0.7150	Enters 2.65 for *S* and solves for *V*.

The horizontal velocity is 0.72 ft/sec.

$$Q = A\text{v} = wd\text{v}$$

$$w = \frac{Q}{d\text{v}} = \frac{\left(3.5\times10^6\ \frac{\text{gal}}{\text{day}}\right)\left(0.134\ \frac{\text{ft}^3}{\text{gal}}\right)}{(4\text{ ft})\left(0.72\ \frac{\text{ft}}{\text{sec}}\right)\left(86{,}400\ \frac{\text{sec}}{\text{day}}\right)}$$

$$= 1.885\text{ ft}\quad(1.9\text{ ft})$$

The answer is (B).

PROBLEM 4

What is most nearly the ultimate bearing capacity for the continuous footing shown? Use the Terzaghi-Meyerhof bearing capacity equation and Terzaghi bearing capacity factors.

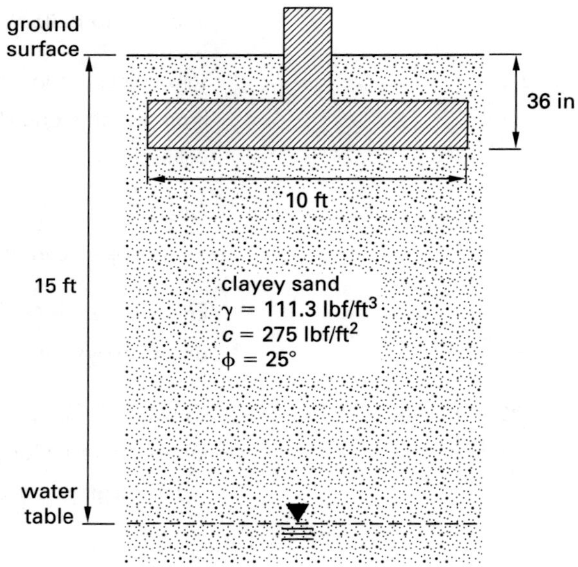

(A) 8300 lbf/ft^2
(B) 14,000 lbf/ft^2
(C) 17,000 lbf/ft^2
(D) 22,000 lbf/ft^2

SOLUTION 4

Use the Terzaghi-Meyerhof equation (Equation 12, p. 22). Terzaghi bearing capacity factors for $\phi = 25°$ are obtained from a published table.

$N_\gamma = 9.7$

$N_c = 25.1$

$N_q = 12.7$

There is no additional surcharge, so p_q is zero. The depth of 36 in must be converted to feet to keep the units consistent.

Keys	Display	Description
[↱] [EQN]	Q=0.5×G×B×N+C×X	Enters Equation mode. Scroll through list to bring needed equation to bottom line of display.
[ENTER]	G? 0.0000	Starts to solve for *Q*. Prompts for *G* and displays current value.
[1] [1] [1] [·] [3] [R/S]	B? 0.0000	Enters 111.3 for *G*. Prompts for *B* and displays current value.
[1] [0] [R/S]	N? 0.0000	Enters 10 for *B*. Prompts for *N* and displays current value.
[9] [·] [7] [R/S]	C? 0.0000	Enters 9.7 for *N*. Prompts for *C* and displays current value.
[2] [7] [5] [R/S]	X? 0.0000	Enters 275 for *C*. Prompts for *X* and displays current value.
[2] [5] [·] [1] [R/S]	P? 0.0000	Enters 25.1 for *X*. Prompts for *P* and displays current value.
[0] [R/S]	D? 0.0000	Enters 0 for *P*. Prompts for *D* and displays current value.
[3] [R/S]	Y? 0.0000	Enters 3 for *D*. Prompts for *Y* and displays current value.

1 2 · 7 R/S	Q = 16,541.0800	Enters 12.7 for *D* and solves for *Q*.

The answer is (C).

PROBLEM 5

What is most nearly the active resultant for the retaining wall shown?

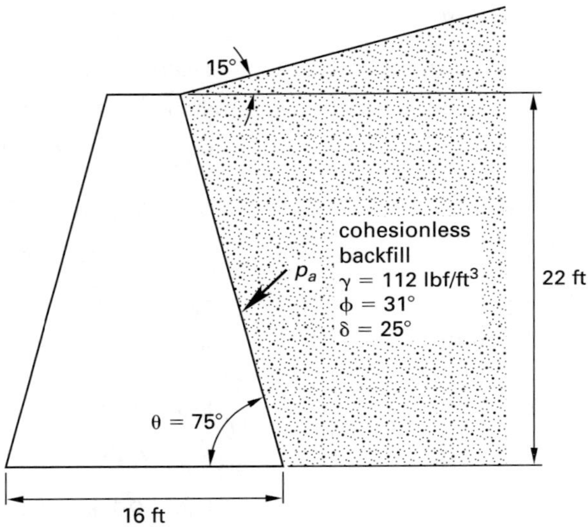

(A) 8700 lbf/ft
(B) 15,000 lbf/ft
(C) 19,000 lbf/ft
(D) 29,000 lbf/ft

SOLUTION 5

Use the formula for the active earth pressure coefficient (Equation 14, p. 25).

Keys	Display	Description
↱ EQN	K=SQ(SIN(T+P))	Enters Equation mode. Scroll through list to bring needed equation to bottom line of display.

[ENTER]	T? 0.0000	Starts to solve for *K*. Prompts for *T* and displays current value.
[7] [5] [R/S]	P? 0.0000	Enters 75 for *T*. Prompts for *P* and displays current value.
[3] [1] [R/S]	D? 0.0000	Enters 31 for *P*. Prompts for *D* and displays current value.
[2] [5] [R/S]	B? 0.0000	Enters 25 for *D*. Prompts for *B* and displays current value.
[1] [5] [R/S]	K = 0.5408	Enters 15 for *B* and solves for *K*.

The coefficient of active earth pressure is 0.54. The active resultant can be calculated.

$$\begin{aligned} p_a &= \frac{1}{2}k_a\gamma H^2 = \left(\frac{1}{2}\right)(0.54)\left(112\ \frac{\text{lbf}}{\text{ft}^2}\right)(22\ \text{ft})^2 \\ &= 14{,}636.16\ \text{lbf/ft} \quad (15{,}000\ \text{lbf/ft}) \end{aligned}$$

The answer is (B).

PROBLEM 6

Using the soil profile data given, what is most nearly the number of days required for 75% consolidation?

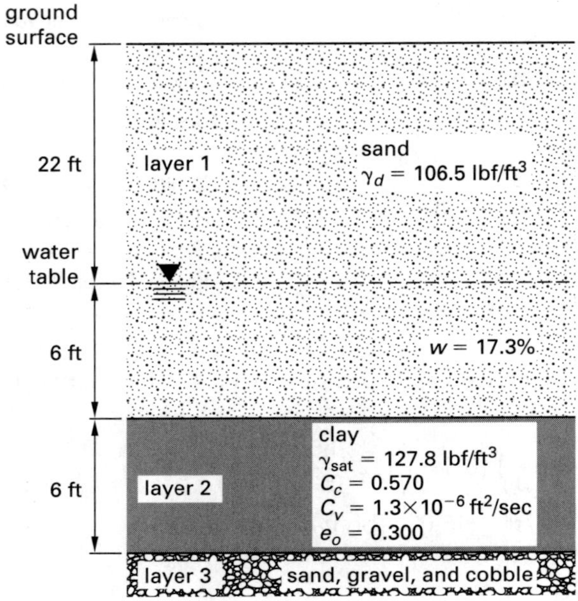

(not to scale)

(A) 15 day
(B) 38 days
(C) 70 days
(D) 160 days

SOLUTION 6

The time factor required for 75% consolidation is typically taken as 0.477. Based on the subsurface soil profile, the clay layer is double-drained with a 3 ft drainage path. Use the equation for the time to reach a specific vertical consolidation (Equation 20, p. 31).

The answer is wanted in days, but C_v is given in square feet per second. Multiply C_v by 86,400 sec/day to convert this rate to square feet per day. This is done within the solution process here, but could also be done before entering Equation mode. Alternately, use Equation mode to solve for the number of seconds needed, and then divide the result by 86,400 to get the number of days.

Keys	Display	Description
[↱] [EQN]	T=V×SQ(H)÷C	Enters Equation mode. Scroll through list to bring needed equation to bottom line of display.
[ENTER]	V? 0.0000	Starts to solve for *T*. Prompts for *V* and displays current value.
[.] [4] [7] [7] [R/S]	H? 0.0000	Enters 0.477 for *V*. Prompts for *H* and displays current value.
[3] [R/S]	C? 0.0000	Enters 3 for *H*. Prompts for *C* and displays current value.
[1] [.] [3] [E] [+/−] [6] [ENTER] [8] [6] [4] [0] [0] [×] [R/S]	T = 38.2212	Converts 1.3×10^{-6} ft^2/sec to square feet per day, enters the result for *C*, and solves for *T*.

The answer is (B).

PROBLEM 7

A circular spiral concrete column supports a 300 kip dead load and a 350 kip live load. The concrete compressive strength is 4000 lbf/in^2, and the yield stress of the reinforcement is 60,000 lbf/in^2. If the maximum reinforcement is used, what is most nearly the cross-sectional area of the column?

(A) 130 in^2
(B) 150 in^2
(C) 200 in^2
(D) 230 in^2

SOLUTION 7

Section 10.3.6 of ACI 318 gives the design axial load strength for compression members.

$$\rho_{g_{max}} = 0.08 \quad \text{[ACI 318 Sec. 10.9.1]}$$

$$\beta = 0.85 \quad \text{[for spiral columns, ACI 318 Sec. 10.3.5.3]}$$

$$\phi = 0.75 \quad \text{[for spiral columns, ACI 318 Sec. 10.3.5.3]}$$

$$\begin{aligned} P_u &= 1.4D + 1.7L \\ &= (1.4)(300 \text{ kips}) + (1.7)(350 \text{ kips}) \\ &= 1015 \text{ kips} \end{aligned}$$

Use the equation for the cross-sectional area of a column (Equation 23, p. 34). Take care to enter U, C, and Y in consistent units, whether kips (as used here) or pounds.

Keys	Display	Description
[⮫] [EQN]	A=U÷P÷B÷(0.85×	Enters Equation mode. Scroll through list to bring needed equation to bottom line of display.
[ENTER]	U? 0.0000	Starts to solve for *A*. Prompts for *U* and displays current value.
[1] [0] [1] [5] [R/S]	P? 0.0000	Enters 1015 for *U*. Prompts for *P* and displays current value.
[·] [7] [5] [R/S]	B? 0.0000	Enters 0.75 for *P*. Prompts for *B* and displays current value.
[·] [8] [5] [R/S]	C? 0.0000	Enters 0.85 for *B*. Prompts for *C* and displays current value.
[4] [R/S]	G? 0.0000	Enters 4 for *C*. Prompts for *G* and displays current value.
[·] [0] [8] [R/S]	Y? 0.0000	Enters 0.08 for *G*. Prompts for *Y* and displays current value.
[6] [0] [R/S]	A = 200.8271	Enters 60 for *Y* and solves for *A*.

The answer is (C).

PROBLEM 8

The W18 × 40 composite steel beam shown supporting a 4 in concrete slab spans 30 ft. The modular ratio is 8.

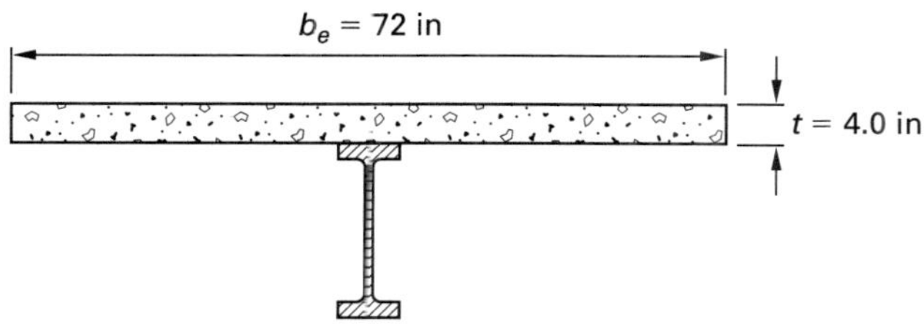

If the concrete slab is transformed to steel, what is the distance from the bottom of the beam to the centroid of the transformed section?

(A) 14 in
(B) 16 in
(C) 17 in
(D) 19 in

SOLUTION 8

Find the section properties of a W18 × 40 beam in Part 1 of the AISC ASD manual.

$b_f = 6.0$ in

$A = 11.8$ in^2

$d = 17.90$ in

To find the centroid of the transformed area, determine the equivalent width of the transformed concrete.

$$b_{\text{transformed}} = \left(\frac{1}{n}\right) b_e = \left(\frac{1}{8}\right)(72 \text{ in})$$
$$= 9.0 \text{ in}$$

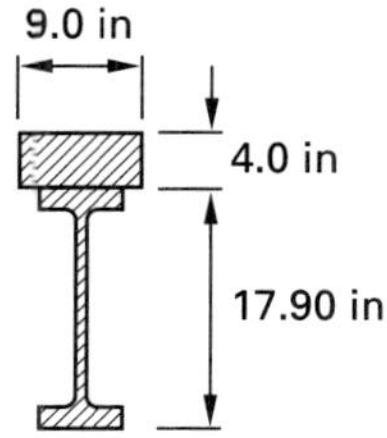

Use the equation for the distance from the neutral axis to the bottom of a steel beam (Equation 26, p. 38). The area of the steel beam is calculated here within the solving process; it could also be calculated before entering Equation mode. Y equals half the thickness of the transformed concrete.

Keys	Display	Description
[↱] [EQN]	B=(S×D÷2+C×(D+	Enters Equation mode. Scroll through list to bring needed equation to bottom line of display.
[ENTER]	S? 0.0000	Starts to solve for *B*. Prompts for *S* and displays current value.
[1] [1] [·] [8] [R/S]	D? 0.0000	Enters 11.8 for *S*. Prompts for *D* and displays current value.
[1] [7] [·] [9] [R/S]	C? 0.0000	Enters 17.9 for *D*. Prompts for *C* and displays current value.
[9] [ENTER] [4] [×] [R/S]	Y? 0.0000	Enters 9 × 4 for *C*. Prompts for *Y* and displays current value.
[2] [R/S]	B = 17.1969	Enters 2 for *Y* and solves for *B*.

The answer is (C).

PROBLEM 9

A 20 in by 24 in short concrete column carries an axial factored load of 750 kips and a factored moment of 600 ft-kips. Bars are located on the short faces of the columns and γ is 0.75. The concrete has a compressive strength of 4000 lbf/in^2 and the reinforcement has a yield stress of 60,000 lbf/in^2. The required area of steel is most nearly

(A) 12 in^2
(B) 14 in^2
(C) 19 in^2
(D) 36 in^2

SOLUTION 9

Interaction diagrams such as those found in the *Civil Engineering Reference Manual* or other sources provide a means to determine the required area of steel when the loads and section properties of a column are known.

Find the x-value (Equation 27, p. 39). As the factored moment is given, use this value for the nominal moment, M, and 1 for the strength reduction factor, P. Take care to use consistent units; inches and kips are used here. The area must be calculated, and the moment of 600 ft-kips must be converted to in-kips. Both are done here within the solving process, but could also be done before entering Equation mode.

Keys	Display	Description
[↱] [EQN]	X=P×M÷A÷H	Enters Equation mode. Scroll through list to bring needed equation to bottom line of display.
[ENTER]	P? 0.0000	Starts to solve for X. Prompts for P and displays current value.
[1] [R/S]	M? 0.0000	Enters 1 for P. Prompts for M and displays current value.
[6] [0] [0] [ENTER] [1] [2] [×] [R/S]	A? 0.0000	Converts 600 ft-kips to in-kips and enters result for M. Prompts for A and displays current value.
[2] [0] [ENTER] [2] [4] [×] [R/S]	H? 0.0000	Enters 20 × 24 for A. Prompts for H and displays current value.
[2] [4] [R/S]	X = 0.6250	Enters 24 for H and solves for X.

Find the y-value (Equation 28, p. 40). As the factored axial load is given, use this value for the nominal load, N, and 1 for the strength reduction factor, P. The previous values of P and A can be retained.

Keys	Display	Description
[⮫] [EQN]	Y=P×N÷A	Enters Equation mode. Scroll through list to bring needed equation to bottom line of display.
[ENTER]	P? 1.0000	Starts to solve for *Y*. Prompts for *P* and displays current value.
[R/S]	N? 0.0000	Retains current value for *P*. Prompts for *N* and displays current value.
[7] [5] [0] [R/S]	A? 480.0000	Enters 750 for *N*. Prompts for *A* and displays current value.
[R/S]	Y = 1.5625	Retains current value for *A* and solves for *Y*.

From interaction diagram E4-60.75 (*Civil Engineering Reference Manual*), using 0.63 ksi and 1.56 ksi as the x- and y-values, respectively, ρ_g is found to be 0.03.

$$\begin{aligned} A_s &= \rho_g A_g = (0.03)(20 \text{ in})(24 \text{ in}) \\ &= 14.4 \text{ in}^2 \quad (14 \text{ in}^2) \end{aligned}$$

The answer is (B).

PROBLEM 10

A equal-tangent crest vertical curve with a length of 400 ft connects grades of +1.0% and −1.75%. The vertex is located at station 35+00 and elevation 549.20 ft. What is most nearly the elevation at sta 35+00?

(A) 543 ft
(B) 546 ft
(C) 548 ft
(D) 549 ft

SOLUTION 10

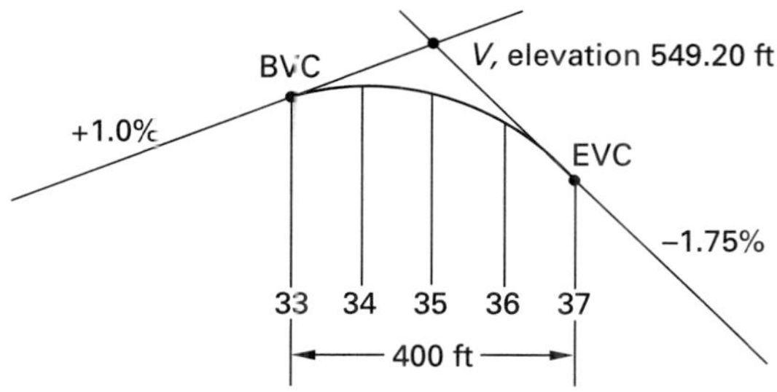

The curve length is 4 sta (400 ft). The beginning of the curve is at station 33+00, and the elevation at this point is

$$z_{\text{BVC}} = z_V - G_1\left(\frac{L}{2}\right) = 549.20 \text{ ft} - \left(1\frac{\text{ft}}{\text{sta}}\right)\left(\frac{4 \text{ sta}}{2}\right)$$
$$= 547.20 \text{ ft}$$

The rate of change is

$$R = \frac{G_2 - G_1}{L} = \frac{-1.75\% - 1\%}{4 \text{ sta}}$$
$$= -0.6875\ \%/\text{sta}$$

Use the formula for elevation at a point on a curve (Equation 32, p. 45).

Keys	Display	Description
[↱] [EQN]	Z=R÷2×SQ(X)+G×	Enters Equation mode. Scroll through list to bring needed equation to bottom line of display.
[ENTER]	R? 0.0000	Starts to solve for *Z*. Prompts for *R* and displays current value.
[·] [6] [8] [7] [5] [+/−] [R/S]	X? 0.0000	Enters −0.6875 for *R*. Prompts for *X* and displays current value.
[2] [R/S]	G? 0.0000	Enters 2 for *X*. Prompts for *G* and displays current value.
[1] [R/S]	B? 0.0000	Enters 1 for *G*. Prompts for *B* and displays current value.

[5] [4] [7] [·] [2] [R/S]	Z = 547.8250	Enters 547.2 for *B* and solves for *Z*.

The answer is (C).

PROBLEM 11

A 2% upgrade meets a 1.5% downgrade on a road with a design speed of 50 mph. The AASHTO recommendation for stopping sight distance is 425 ft. What is most nearly the minimum length of vertical curve required?

(A) 233 ft
(B) 299 ft
(C) 360 ft
(D) 425 ft

SOLUTION 11

The change in gradient is

$$A = |G_2 - G_1| = |(-1.5\%) - 2.0\%| = 3.5\%$$

Use the equations for a crest vertical curve, using 3.5 ft for G and 2.0 ft for H. Assume $S < L$ (Equation 33, p. 46).

Keys	**Display**	**Description**
[↱] [EQN]	L=A×SQ(S)÷200÷	Enters Equation mode. Scroll through list to bring needed equation to bottom line of display.
[ENTER]	A? 0.0000	Starts to solve for *L*. Prompts for *A* and displays current value.
[3] [·] [5] [R/S]	S? 0.0000	Enters 3.5 for *A*. Prompts for *S* and displays current value.
[4] [2] [5] [R/S]	G? 0.0000	Enters 425 for *S*. Prompts for *G* and displays current value.
[3] [·] [5] [R/S]	H? 0.0000	Enters 3.5 for *G*. Prompts for *H* and displays current value.

[2] [R/S]	L = 292.9099	Enters 2 for *H* and solves for *L*.

If $S < L$, then L is 293 ft. S is 425 ft and greater than L, so this is not a valid answer.

Assume $S > L$ (Equation 34, p. 47). The values are the same as before and need not be re-entered.

Keys	**Display**	**Description**
[↱] [EQN]	L=2×S-(200×SQ(	Enters Equation mode. Scroll through list to bring needed equation to bottom line of display.
[ENTER]	S? 425.0000	Starts to solve for *L*. Prompts for *S* and displays current value.
[R/S]	G? 3.5000	Retains current value for *S*. Prompts for *G* and displays current value.
[R/S]	H? 2.0000	Retains current value for *G*. Prompts for *H* and displays current value.
[R/S]	A? 0.0000	Retains current value for *H*. Prompts for *A* and displays current value.
[R/S]	L = 233.3427	Retains current value for *A* and solves for *L*.

If $S > L$, then L is 233 ft. S is 425 ft and greater than L, so this is a valid answer. The minimum length is 233 ft.

The answer is (A).

PROBLEM 12

A 1.25% downgrade meets a 3.0% upgrade on a road with a design speed of 60 kph. The AASHTO recommendation for stopping sight distance is 85 m. What is most nearly the minimum length of vertical curve required?

(A) 72 m
(B) 77 m
(C) 85 m
(D) 88 m

SOLUTION 12

The change in gradient is

$$A = |G_2 - G_1| = |3.0\% - (-1.25\%)|$$
$$= 4.25\%$$

Use the equations for a sag vertical curve. Assume $S < L$ (Equation 35, p 48). Because the problem is in SI units, use 120 for X.

Keys	Display	Description
[↱] [EQN]	`L=A×SQ(S)÷(X+3.`	Enters Equation mode. Scroll through list to bring needed equation to bottom line of display.
[ENTER]	`A?` `0.0000`	Starts to solve for *L*. Prompts for *A* and displays current value.
[4] [·] [2] [5] [R/S]	`S?` `0.0000`	Enters 4.25 for *A*. Prompts for *S* and displays current value.
[8] [5] [R/S]	`X?` `0.0000`	Enters 85 for *S*. Prompts for *X* and displays current value.
[1] [2] [0] [R/S]	`L =` `73.5479`	Enters 120 for *X* and solves for *L*.

If $S < L$, the minimum length is 74 m. S is 85 m and greater than L, so this is not a valid answer.

Assume $S > L$ (Equation 36, p. 50). The values are the same as before and need not be re-entered.

Keys	Display	Description
[↱] [EQN]	`L=2×S-(X+3.5×S)`	Enters Equation mode. Scroll through list to bring needed equation to bottom line of display.

[ENTER]	S? 85.0000	Starts to solve for *L*. Prompts for *S* and displays current value.
[R/S]	X 120.0000	Retains current value for *S*. Prompts for *X* and displays current value.
[R/S]	A? 4.2500	Retains current value for *X*. Prompts for *A* and displays current value.
[R/S]	L = 71.7647	Retains current value for *A* and solves for *L*.

If $S > L$, then L is 72 m. S is 85 m and greater than L, so this is a valid answer. The minimum length is 72 m.

The answer is (A).

PROBLEM 13

What is most nearly the flow rate through the venturi meter shown? Water and mercury are both at 70°F. Assume the velocity coefficient is 1.0.

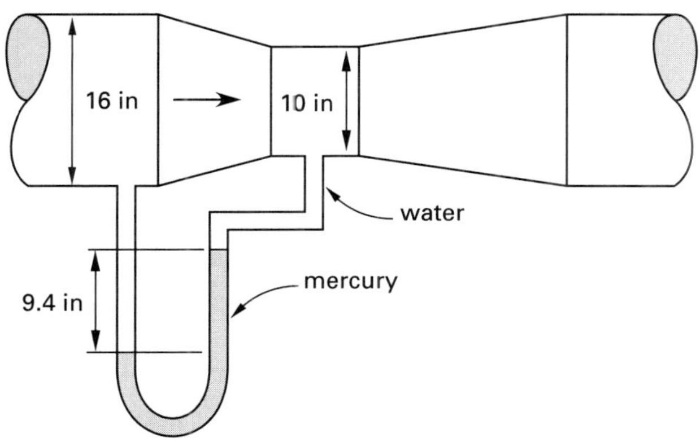

(A) 5.9 ft^3/sec
(B) 15 ft^3/sec
(C) 23 ft^3/sec
(D) 38 ft^3/sec

SOLUTION 13

Use the formula for throat velocity in a venturi meter (Equation 50, p. 68). The approximate density of mercury at room temperature is 848 lbm/ft^3, and that of water is 62.4 lbm/ft^3. The beta ratio for the venturi meter is

$$\beta = \frac{d_2}{d_1} = \frac{10 \text{ in}}{16 \text{ in}}$$
$$= 0.625$$

To keep units consistent, the manometer reading must be converted from inches to feet, by dividing by 12 in/ft. This is done here within the solving process, but could also be done before entering Equation mode.

Keys	Display	Description
[↱] [EQN]	`V=C÷SQRT(1-B^4`	Enters Equation mode. Scroll through list to bring needed equation to bottom line of display.
[ENTER]	`C?` `0.0000`	Starts to solve for *V*. Prompts for *C* and displays current value.
[1] [R/S]	`B?` `0.0000`	Enters 1 for *C*. Prompts for *B* and displays current value.
[·] [6] [2] [5] [R/S]	`G?` `0.0000`	Enters 0.625 for *B*. Prompts for *G* and displays current value.
[3] [2] [·] [2] [R/S]	`M?` `0.0000`	Enters 32.2 for *G*. Prompts for *M* and displays current value.
[8] [4] [8] [R/S]	`P?` `0.0000`	Enters 848 for *M*. Prompts for *P* and displays current value.
[6] [2] [·] [4] [R/S]	`H?` `0.0000`	Enters 62.4 for *P*. Prompts for *H* and displays current value.
[9] [·] [4] [ENTER] [1] [2] [÷] [R/S]	`V =` `27.3765`	Converts 9.4 in to feet, enters 0.7833 for *H*, and solves for *V*.

The throat velocity is 27.4 ft/sec. Multiply by the cross-sectional area of the throat to get the flow rate.

$$A_2 = \frac{\pi d_2^2}{4} = \left(\frac{\pi\,(10\text{ in})^2}{4}\right)\left(\frac{1\text{ ft}}{12\text{ in}}\right)^2$$
$$= 0.545\text{ ft}^2$$
$$Q = A_2 v_2 = \left(0.545\text{ ft}^2\right)\left(27.4\ \frac{\text{ft}}{\text{sec}}\right)$$
$$= 14.9\text{ ft}^3/sec$$

The answer is (B).

PROBLEM 14

A run of schedule-40 pipe with long radius 90° elbows is fixed in an overhead rack. The interior diameter of the pipe is 3 in. The elbows are in the horizontal plane with no change in inlet and outlet elevations. The flow in the pipe is 0.55 ft^3/sec at 50 psig. What is most nearly the resultant force exerted by the water on the pipe rack at each elbow?

(A) 520 lbf
(B) 540 lbf
(C) 650 lbf
(D) 730 lbf

SOLUTION 14

The approximate density of water at room temperature is 62.4 lbm/ft^3. The mass flow rate is

$$\dot{m} = Q\rho = \left(0.55\ \frac{\text{ft}^3}{\text{sec}}\right)\left(62.4\ \frac{\text{lbm}}{\text{ft}^3}\right)$$
$$= 34.3\text{ lbm/sec}$$

The cross-sectional area of the pipe in feet is

$$A = \frac{\pi d^2}{4} = \frac{\pi\,(3\text{ in})^2}{4}$$
$$= \left(7.07\text{ in}^2\right)\left(\frac{1\text{ ft}}{12\text{ in}}\right)^2$$
$$= 0.049\text{ ft}^2$$

The flow velocity is

$$\text{v} = \frac{Q}{A} = \frac{0.55\ \frac{\text{ft}^3}{\text{sec}}}{0.049\ \text{ft}^2} = 11.2\ \text{ft/sec}$$

Because the elbows have a constant diameter, the pressure and area are constant. To keep units consistent, pressure must be converted from pounds per square inch to pounds per square foot by multiplying by 144 in^2/ft^2. This is done here within the solving process, but could also be done before entering Equation mode. Use the formulas for a confined stream in a pipe bend (Equations 46-47, p. 63).

Keys	Display	Description
[⮫] [EQN]	`X=P×A×COS(T)-O`	Enters Equation mode. Scroll through list to bring needed equation to bottom line of display.
[ENTER]	`P?` `0.0000`	Starts to solve for *X*. Prompts for *P* and displays current value.
[5] [0] [ENTER] [1] [4] [4] [×] [R/S]	`A?` `0.0000`	Converts 50 psi to pounds per square feet and enters 7200 for *P*. Prompts for *A* and displays current value.
[·] [0] [4] [9] [R/S]	`T?` `0.0000`	Enters 0.049 for *A*. Prompts for *T* and displays current value.
[9] [0] [R/S]	`O?` `0.0000`	Enters 90 for *T*. Prompts for *O* and displays current value.
[5] [0] [ENTER] [1] [4] [4] [×] [R/S]	`Z?` `0.0000`	Converts 50 psi to pounds per square feet and enters 7200 for *O*. Prompts for *Z* and displays current value.
[·] [0] [4] [9] [R/S]	`M?` `0.0000`	Enters 0.049 for *Z*. Prompts for *M* and displays current value.
[3] [4] [·] [3] [R/S]	`V?` `0.0000`	Enters 34.3 for *M*. Prompts for *V* and displays current value.

1 1 . 2 R/S	U? 0.0000	Enters 11.2 for *V*. Prompts for *U* and displays current value.
1 1 . 2 R/S	G? 0.0000	Enters 11.2 for *U*. Prompts for *G* and displays current value.
3 2 . 2 R/S	X = −364.7304	Enters 32.2 for *G*, and solves for *X*.

The x-component of the force is −365 lbf. Values in the second equation are the same as before and need not be re-entered.

Keys	**Display**	**Description**
↱ EQN	Y=(P×A+M×V÷G)×	Enters Equation mode. Scroll through list to bring needed equation to bottom line of display.
ENTER	P? 7,200.0000	Starts to solve for *X*. Prompts for *P* and displays current value.
R/S	A? 0.0490	Retains current value for *P*. Prompts for *A* and displays current value.
R/S	M? 34.3000	Retains current value for *A*. Prompts for *M* and displays current value.
R/S	V? 11.2000	Retains current value for *M*. Prompts for *V* and displays current value.
R/S	G? 0.0000	Retains current value for *V*. Prompts for *G* and displays current value.
R/S	T? 90.0000	Retains current value for *G*. Prompts for *T* and displays current value.
R/S	Y = 364.7304	Retains current value for *T* and solves for *Y*.

The y-component of the force is 365 lbf. The resultant force is

$$F_R = \sqrt{F_x^2 + F_y^2} = \sqrt{(-365\ \text{lbf})^2 + (365\ \text{lbf})^2}$$
$$= 516\ \text{lbf} \quad (520\ \text{lbf})$$

The answer is (A).

PROBLEM 15

A submersible pump operated at 1750 rpm must be used to deliver 800 gpm of water from a 350 ft deep well. At 93% efficiency, the average specific speed of the pump is 2300. Under these conditions, how many stages are required for the pump?

(A) 1 stage
(B) 3 stages
(C) 6 stages
(D) 35 stages

SOLUTION 15

Use the equation for the specific speed of a pump (Equation 44, p. 60) to find the head added by the pump with a single stage.

Keys	Display	Description
[↱] [EQN]	`S=N×SQRT(Q)÷H^`	Enters Equation mode. Scroll through list to bring needed equation to bottom line of display.
[SOLVE] [H]	`S?` `0.0000`	Starts to solve for *H*. Prompts for *S* and displays current value.
[2] [3] [0] [0] [R/S]	`N?` `0.0000`	Enters 2300 for *S*. Prompts for *N* and displays current value.
[1] [7] [5] [0] [R/S]	`Q?` `0.0000`	Enters 1750 for *N*. Prompts for *Q* and displays current value.
[8] [0] [0] [R/S]	`H =` `59.8605`	Enters 800 for *Q* and solves for *H*.

The head added by the pump with one stage is 59.9 ft. To achieve 350 ft of head, the number of stages needed must be at least 350 ft/59.9 ft, or 5.84. Thus 6 stages are required.

The answer is (C).

PRACTICE PROBLEMS USING HP 33s BUILT-IN FUNCTIONS

C

The equations in Part A and the sample problems in Part B make use of most of the preprogrammed functions on the hp33s that you're likely to use during the civil PE examination. These include exponential, logarithmic, power, and trigonometric functions.

Part C demonstrates some additional functions that may be useful, including probability and statistics, manipulating complex numbers, and converting between number systems.

The hp33s contains other preprogrammed functions, such as the quotient and remainder, percentage, and numerical functions (integer part, absolute value, and so on), but these are much less likely to be of use on the exam.

COMPLEX NUMBERS: MULTIPLICATION

What is the product of the complex numbers $3 + 4i$ and $7 - 2i$?

(A) $10 + 2i$
(B) $13 + 22i$
(C) $13 + 34i$
(D) $29 + 22i$

SOLUTION

Keys	Display	Description
[4] [ENTER]	4.0000 4.0000	Enters the first complex number. In RPN mode, the *y*-value is entered first.
[3] [ENTER]	3.0000 3.0000	Enters the *x*-value.

Keys	Display	Description
[2] [+/−] [ENTER]	−2.0000 −2.0000	Enters the second complex number. In RPN mode, the *y*-value is entered first.
[7]	−2.0000 7	Enters the *x*-value.
[⬏] [CMPLX] [×]	22.0000 29.0000	Selects complex multiplication. The *x*-value of the answer appears in the X-register, and the *y*-value in the Y-register.

The answer is (D).

COMPLEX NUMBERS: CONVERSION FROM RECTANGULAR TO POLAR COORDINATES

What is the exponential form of the complex number $3 + 4i$?

(A) $e^{i53.1^\circ}$
(B) $5e^{i53.1^\circ}$
(C) $5e^{i126.9^\circ}$
(D) $7e^{i53.1^\circ}$

SOLUTION

The answer will be in the form $re^{i\theta}$.

Keys	Display	Description
[4] [ENTER]	4.0000 4.0000	Starts to enter the complex number. Enter the *y*-value first.
[3]	3.0000 3	Enters the *x*-value.
[⬏] [→θ,r]	53.1301 5.0000	Selects conversion to polar coordinates. The *r*-value of the answer appears in the X-register, and the θ-value in the Y-register.

The answer is (B).

STATISTICS: MEAN AND SAMPLE STANDARD DEVIATION

What are most nearly the mean and sample standard deviation of the following numbers?

71.3, 74.0, 74.25, 78.54, 80.6

(A) 74.3, 2.7
(B) 74.3, 3.7
(C) 75.0, 2.7
(D) 75.7, 3.8

SOLUTION

Keys	Display	Description
[left-shift] [CLEAR]	1X 2VARS 3ALL 4Σ	
[4]	0.0000	Selects Σ to clear statistics registers.
[7] [1] [.] [3] [Σ+]	1.0000	Enters first of five data points.
[7] [4] [.] [0] [Σ+]	2.0000	
[7] [4] [.] [2] [5] [Σ+]	3.0000	
[7] [8] [.] [5] [4] [Σ+]	4.0000	
[8] [0] [.] [6] [Σ+]	5.0000	
[right-shift] [$\bar{x},\bar{y}$]	$\bar{x}$ $\bar{y}$ $\bar{x}$W 75.7380	Determines mean.
[right-shift] [S,σ]	sx sy σx σy 3.7557	Determines standard deviation for same data.

The answer is (D).

Selecting σx will give the population standard deviation for the data.

When two-variable data have been entered, selecting ȳ will give the mean of the y-values, and selecting x̄W will give the mean of the x-values using the y-values as weights. Selecting sy and σy will give the sample standard deviation and the population standard deviation, respectively, for the y-values.

PROBABILITY: COMBINATIONS

In how many ways can a 5-person committee be formed within an organization with 14 members?

(A) 700
(B) 1365
(C) 2002
(D) 5040

SOLUTION

Find the number of combinations from a set of n items, taken r at a time.

Keys	Display	Description
[1] [4]	14.0000	Enters values for *n* and *r*.
[ENTER] [5]	5	
[↰] [nCr]	2,002.0000	Finds the number of combinations.

The answer is (C).

COMPUTER MATHEMATICS: BASE CONVERSIONS

Convert the hexadecimal number 4D3 to base-10.

(A) 1219
(B) 1222
(C) 1235
(D) 1251

SOLUTION

Keys	Display	Description
[↰] [BASE]	1DEC 2HEX 3OCT 4BIN	Allows choice of base.

[2]	0	Selects hexadecimal. **HEX** annunciator appears.
[4] [D] [3]	4D3_	Enters 4D3.
[↰] [BASE]	1DEC 2HEX 3OCT 4BIN	
[1]	1,235.0000	Selects decimal. **HEX** annunciator disappears.

The answer is (C).

PRACTICE PROBLEMS USING ALGEBRAIC MODE

D

Most HP owners use their calculators in Reverse Polish Notation or RPN mode. This mode uses less memory and tends to work faster. But some prefer Algebraic or ALG mode, which can be easier to read and write.

In ALG mode, certain functions and procedures are different. Most of the hp33s User's Manual shows RPN mode, but a checkmark is shown next to any step performed differently in ALG mode. Appendix C of the User's Manual also summarizes the differences.

This section shows a few of these differences in action, focusing on situations most likely to come up on the civil PE examination.

The most common difference between the two modes is in two-number operations. In RPN mode, the ENTER key is used to separate the two numbers, and the operator key is pressed last, causing the calculation to occur. In ALG mode, this is reversed. The operator key separates the two numbers, and the ENTER key triggers the calculation. For example, to raise three to the fifth power in ALG mode, you enter [3] [y^x] [5] [ENTER], whereas in RPN mode the sequence is [3] [ENTER] [5] [y^x].

In ALG mode, parentheses may be used up to 13 levels deep. The operation of the stack is different, but this is important only in programming. In RPN mode, two-variable statistical data is entered with the y-value first; in ALG mode, the x-value is entered first.

Finally, complex number arithmetic and coordinate conversions are somewhat different in the two modes. The sample problems in this section illustrate some of the differences. Each was solved in Part C using RPN mode; in this section the same problems are solved using ALG mode.

COMPLEX NUMBERS: MULTIPLICATION

What is the product of the complex numbers $3+4i$ and $7-2i$?

(A) $10+2i$
(B) $13+22i$
(C) $13+34i$
(D) $29+22i$

SOLUTION

Keys	Display	Description
[↱] [ALG]	0.0000	Shifts to Algebraic mode. **ALG** annunciator appears.
[↱] [(] [3] [+]	(3+ 3.0000	Enters real portion of first number.
[4] [↰] [CMPLX] [↱] [)]	(3+4i) RE=3.0000	Enters imaginary portion. In Algebraic mode, the [CMPLX] key is used after the imaginary part of a complex number. RE indicates the real portion; if desired, scroll down with the silver key to display the imaginary portion.
[×]	(3+4i)× RE=3.0000	This step is optional. In Algebraic mode, implied multiplication is allowed.
[↱] [(] [7] [−] [2] [↰] [CMPLX] [↱] [)]	(3+4i)×(7−2i) RE=7.0000	Enters second complex number.
[ENTER]	(3+4i)×(7−2i) RE=29.0000	Performs calculation and displays real portion of answer.
	(3+4i)×(7−2i) IM=22.0000	Scroll down with large silver key to display imaginary portion.

The answer is (D).

COMPLEX NUMBERS: CONVERSION FROM RECTANGULAR TO POLAR COORDINATES

What is the exponential form of the complex number $3 + 4i$?

(A) $e^{i53.1°}$
(B) $5e^{i53.1°}$
(C) $5e^{i126.9°}$
(D) $7e^{i53.1°}$

SOLUTION

The answer will be in the form $re^{i\theta}$.

Keys	Display	Description
[↱] [ALG]	0.0000	Shifts to Algebraic mode. **ALG** annunciator appears.
[4]	4_	Starts to enter complex number. Enter *y*-value first.
[x◂▸y]	0.0000	Exchanges X- and Y-registers.
[3]	3_	Enters *x*-value.
[↰] [→θ,r]	4,3→θ,r r=5.0000	Selects conversion to polar coordinates and displays *r*-value of answer.
	4,3→θ,r θ=53.1301	Scroll down with large silver key to display θ-value of the answer

The answer is (B).

PROBABILITY: COMBINATIONS

In how many ways can a 5-person committee be formed within an organization with 14 members?

(A) 700
(B) 1365
(C) 2002
(D) 5040

SOLUTION

Find the number of combinations from a set of n items, taken r at a time.

Keys	Display	Description
[↱] [ALG]	0.0000	Shifts to algebraic mode. **ALG** annunciator appears.
[1] [4] [↰] [nCr]	14nCr 14.0000	Enters value for *n*.
[5] [ENTER]	14nCr5= 2,002.0000	Enters value for *r* and finds number of combinations.

The answer is (C).

QUICK REFERENCE

E

Equation 1: Mass Balance

$$C_f = \frac{C_1Q_1 + C_2Q_2}{Q_1 + Q_2}$$

Entered as

```
F=(B×P+C×Q)÷(P+Q)
```

Variables

$F = C_f$	final concentration	mg/gal	mg/L
$B = C_1$	concentration	mg/gal	mg/L
$C = C_2$	concentration	mg/gal	mg/L
$P = Q_1$	flow quantity	gal/day	L/d
$Q = Q_2$	flow quantity	gal/day	L/d

Equation 2: Mean Cell Residence Time

$$\theta_c = \frac{V_aX}{Q_eX_e + Q_wX_w}$$

Entered as

```
T=V×X÷(P×E+Q×W)
```

Variables

$T = \theta_c$	mean cell residence time	day	d
$V = V_a$	volume of aeration tank	ft^3	m^3
$X = X$	mixed liquor volatile suspended solids	mg/L	mg/L
$P = Q_e$	flow rate, effluent	ft^3/day	m^3/d
$Q = Q_w$	flow rate, wasted	ft^3/day	m^3/d
$E = X_e$	mixed liquor volatile suspended solids, effluent	mg/L	mg/L
$W = X_w$	mixed liquor volatile suspended solids, wasted	mg/L	mg/L

Equation 3: Ultimate Biochemical Oxygen Demand

$$\text{BOD}_t = \text{BOD}_u \left(1 - 10^{-K_d t}\right)$$

Entered as

```
T=U×(1-10^(-K×S))
```

Variables

$T = \text{BOD}_t$	biochemical oxygen demand at time t	mg/L
$U = \text{BOD}_u$	ultimate biochemical oxygen demand	mg/L
$K = K_d$	rate constant	d^{-1}
$S = t$	time	d

Equation 4: Population Equivalent

$$P_{\text{eq,1000s}} = \frac{\left(\text{BOD}_{\text{mg/L}}\right)\left(Q_{\text{ML/day}}\right)}{90\ \dfrac{\text{g}}{\text{person-day}}}$$

Entered as

```
P=B×Q÷90
```

Variables

$P = P_{\text{eq,1000s}}$	population in thousands	–
$B = \text{BOD}_{\text{mg/L}}$	biochemical oxygen demand	mg/L
$Q = Q_{\text{ML/day}}$	flow quantity	ML/d

Equation 5: Time to Critical Point in Stream Purification

$$t_c = \left(\frac{1}{K_r - K_d}\right) \log_{10}\left(\left(\frac{K_d\text{BOD}_u - K_r D_0 + K_d D_0}{K_d\text{BOD}_u}\right)\left(\frac{K_r}{K_d}\right)\right)$$

Entered as

```
T=LOG((D×U-R×O+D×O)÷D÷U×R÷D)÷(R-D)
```

Variables

$T = t_c$	time	d
$D = K_d$	rate constant, deoxygenated	d^{-1}
$U = \text{BOD}_u$	ultimate biochemical oxygen demand	mg/L
$R = K_r$	rate constant, reoxygenated	d^{-1}
$O = D_0$	initial oxygen deficit	mg/L

Equation 6: Velocity of Settling

$$v = \sqrt{8k\left(\frac{gd_p}{f}\right)(SG_p - 1)}$$

Entered as

```
V=SQRT(8×K×G×D÷F×(S-1))
```

Variables

$V = v$	velocity	ft/sec	m/s
$K = k$	Camp constant	–	–
$G = g$	gravitational acceleration	32.2 ft/sec^2	9.81 m/s^2
$D = d_p$	diameter of particle	ft	m
$F = f$	Darcy friction factor	–	–
$S = SG_p$	specific gravity	–	–

Equation 7: Temperature Correction

$$K_{r,T_1} = K_{r,T_2}\theta_r^{T_1 - T_2}$$

Entered as

```
J=K×R^(S-T)
```

Variables

$J = K_{r,T_1}$	rate constant	day^{-1}	d^{-1}
$K = K_{r,T_2}$	rate constant	day^{-1}	d^{-1}
$R = \theta_r$	temperature constant	–	–
$S = T_1$	temperature	°F	°C
$T = T_2$	temperature	°F	°C

Equation 8: Specific Substrate Utilization Rate

$$U = \frac{S_o - S}{\theta X}$$

Entered as

```
U=(O-S)÷T÷X
```

Variables

$U = U$	specific substrate utilization	d^{-1}
$O = S_o$	growth-limited substrate concentration, influent	mg/L
$S = S$	growth-limited substrate concentration	mg/L
$T = \theta$	residence/detention time	d
$X = X$	mixed liquor volatile suspended solids	mg/L

Equation 9: Relationship Between Mean Cell Residence Time and Specific Utilization Rate

$$\frac{1}{\theta_c} = YU - k_d$$

Entered as

```
1÷T=Y×U-K
```

Variables

$T = \theta_c$	mean cell residence time	d
$Y = Y$	yield coefficient	mg/mg
$U = U$	specific substrate utilization	d^{-1}
$K = k_d$	microorganism endogenous decay rate	d^{-1}

Equation 10: Observed Yield

$$Y_{obs} = \frac{Y}{1 + k_d\theta_c}$$

Entered as

```
O=Y÷(1+K×T)
```

Variables

$O = Y_{obs}$	observed yield coefficient	mg/mg
$Y = Y$	yield coefficient	mg/mg
$K = k_d$	microorganism endogenous decay rate	d^{-1}
$T = \theta_c$	cell residence time	d

Equation 11: Mohr-Coulomb Equation

$$s = c + \sigma \tan \phi$$

Entered as

```
S=C+N×TAN(P)
```

Variables

$S = s$	shear strength	lbf/ft^2	Pa
$C = c$	cohesion	lbf/ft^2	Pa
$N = \sigma$	normal stress	lbf/ft^2	Pa
$P = \phi$	angle of internal friction	deg	deg

Equation 12: Terzaghi-Meyerhof Equation for Continuous Footing

$$q_{\text{ult}} = \tfrac{1}{2}\gamma B N_\gamma + c N_c + (p_q + \gamma D_f) N_q$$

Entered as

```
Q=0.5×G×B×N+C×X+(P+G×D)×Y
```

Variables

$Q = q_{\text{ult}}$	ultimate bearing capacity	lbf/ft^2	Pa
$G = \gamma$	specific weight	lbf/ft^3	N/m^3
$B = B$	width	ft	m
$N = N_\gamma$	bearing capacity factor, density	–	–
$C = c$	cohesion	lbf/ft^2	Pa
$X = N_c$	bearing capacity factor, cohesion	–	–
$P = p_q$	surcharge pressure	lbf/ft^2	Pa
$D = D_f$	depth of footing	ft	m
$Y = N_q$	bearing capacity factor, surcharge	–	–

Equation 13: Primary Consolidation in Normally Consolidated Clay

$$S_{\text{primary}} = \frac{H C_c \log_{10}\left(\dfrac{p_o' + \Delta p_v'}{p_o'}\right)}{1 + e_o}$$

Entered as

```
S=H×C×LOG((O+V)÷O)÷(1+E)
```

Variables

$S = S_{\text{primary}}$	settlement	ft	m
$H = H$	layer thickness or depth	ft	m
$C = C_c$	compression index	–	–
$O = p_o'$	original effective pressure	lbf/ft^2	Pa
$V = p_v'$	vertical effective pressure	lbf/ft^2	Pa
$E = e_o$	void ratio	–	–

Equation 14: Active Earth Pressure Coefficient (Coulomb Equation)

$$k_a = \frac{\sin^2(\theta + \phi)}{\sin^2\theta \sin(\theta - \delta)\left(1 + \sqrt{\dfrac{\sin(\phi + \delta)\sin(\phi - \beta)}{\sin(\theta - \delta)\sin(\theta + \beta)}}\right)^2}$$

Entered as

```
K=SQ(SIN(T+P))÷SQ(SIN(T))÷SIN(T-D)÷SQ(1+SQRT(SIN(P+D)
    ×SIN(P-B)÷SIN(T-D)÷SIN(T+B)))
```

Variables

$K = k_a$	active earth pressure coefficient	–	–
$T = \theta$	angle of resultant force	deg	deg
$P = \phi$	angle of internal friction	deg	deg
$D = \delta$	angle of external friction	deg	deg
$B = \beta$	slope of backfill	deg	deg

Equation 15: Piles: Point Capacity for Cohesive Soil

$$Q_p \approx 9A_pc$$

Entered as

```
Q=9×A×C
```

Variables

$Q = Q_p$	bearing capacity, pile tip	lbf	N
$A = A_p$	area, pile tip	ft^2	m^2
$C = c$	cohesion	lbf/ft^2	Pa

Equation 16: Piles: Skin-Friction Capacity for Cohesive Soil

$$Q_f = pf_sL_e$$

Entered as

```
Q=P×F×L
```

Variables

$Q = Q_f$	skin-friction capacity	lbf	N
$P = p$	perimeter	ft	m
$F = f_s$	skin friction factor	–	–
$L = L_e$	effective length	ft	m

Equation 17: Piles: Skin-Friction Capacity for Cohesive Soil (β-Method)

$$Q_f = p\beta\sigma' L_e$$

Entered as

```
Q=P×B×S×L
```

Variables

$Q = Q_f$	skin-friction capacity	lbf	N
$P = p$	perimeter	ft	m
$B = \beta$	effective stress factor	–	–
$S = \sigma'$	earth pressure	–	–
$L = L_e$	effective length	ft	m

Equation 18: Increase in Applied Stress Below Loaded Rectangular Area

$$\Delta\sigma_z = q\frac{BL}{(B+z)(L+z)}$$

Entered as

```
S=Q×B×L÷(B+Z)÷(L+Z)
```

Variables

$S = \sigma_z$	stress	lbf/ft^2	Pa
$Q = q$	uniform surcharge	lbf/ft^2	Pa
$B = B$	footing width	ft	m
$L = L$	distance	ft	m
$Z = z$	depth	ft	m

Equation 19: Safe Load for Double-Acting Steam Hammer

$$Q_{a,\text{lbf}} = \frac{2E_{\text{ft-lbf}}}{S_{\text{in}} + 0.1}$$

Entered as

```
Q=2×E÷(S+0.1)
```

Variables

$Q = Q_{a,\text{lbf}}$	allowable vertical load per pile	lbf	N
$E = E_{\text{ft-lbf}}$	energy per blow	ft-lbf	J
$S = S_{\text{in}}$	average penetration per blow	in	m

Equation 20: Time to Reach Specific Vertical Consolidation

$$t = \frac{T_v H_d^2}{C_v}$$

Entered as

```
T=V×SQ(H)÷C
```

Variables

$T = t$	time	sec	s
$V = T_v$	time factor	–	–
$H = H_d$	thickness of layer	ft	m
$C = C_v$	coefficient of consolidation	ft^2/sec	m^2/s

Notes H is full layer thickness for single drainage, half for double drainage. If the coefficient of consolidation is given in units of feet or meters per day, time will be in units of days.

Equation 21: Required Plate Thickness

$$t = 2c\sqrt{\frac{f_p}{F_y}}$$

(c = larger of m and n)

Entered as

```
T=2×C×SQRT(P÷Y)
```

Variables

$T = t$	thickness	in
$C = c$	dimension	in
$P = f_p$	bearing stress	$kips/in^2$
$Y = F_y$	yield strength	$kips/in^2$

Equation 22: Deflection for Simply Supported Beam

$$\delta = \frac{5wL^4}{384EI}$$

Entered as

```
D=5×W×L^4÷384÷E÷I
```

Variables

$D = \delta$	deformation	in	m
$W = w$	load per unit length	lbf/in	N/m
$L = L$	length	in	m
$E = E$	modulus of elasticity	lbf/in^2	Pa
$I = I$	moment of inertia	in^4	m^4

Equation 23: Cross-Sectional Area of Circular Spiral Concrete Column

$$A_g = \frac{P_u}{\phi\beta\left(0.85f'_c\left(1-\rho_g\right)+\rho_g f_y\right)}$$

Entered as

```
A=U÷P÷B÷(0.85×C×(1-G)+G×Y)
```

Variables

$A = A_g$	gross area	in^2	m^2
$U = P_u$	factored axial load	lbf	N
$P = \phi$	strength reduction factor	–	–
$B = \beta$	column strength factor	–	–
$C = f'_c$	compressive strength	lbf/in^2	Pa
$G = p_g$	longitudinal reinforcement ratio	–	–
$Y = f_y$	yield stress	lbf/in^2	Pa

Equation 24: Shear Strength Provided by Concrete

$$V_c = 2\sqrt{f'_c}b_w d$$

Entered as

```
V=2×SQRT(F)×B×D
```

Variables

$V = V_c$	shear strength, concrete	lbf
$F = f'_c$	compressive strength	lbf/in^2
$B = b_w$	width of compressive member, web	in
$D = d$	beam depth (to tension steel centroid)	in

Equation 25: Area of Steel in Reinforced Concrete Beam

$$A_s = \frac{M_u}{\phi f_y\left(d-\lambda\right)}$$

Entered as `A=M÷P÷F÷(D-L)`

Variables

$A = A_s$	area	in^2
$M = M_u$	ultimate moment	in-kips
$P = \phi$	strength reduction factor	–
$F = f_y$	yield stress	ksi
$D = d$	depth	in
$L = \lambda$	distance from centroid of compressed area to extreme compression fiber	in

Equation 26: Distance From Neutral Axis to Bottom of Steel Beam

$$\bar{y}_b = \frac{A_s\left(\frac{d}{2}\right) + A_{\text{ctr}}\,(d + Y2)}{A_s + A_{\text{ctr}}}$$

Entered as

```
B=(S×D÷2+C×(D+Y))÷(S+C)
```

Variables

$B = \overline{y}_b$	distance from neutral axis of composite beam to bottom of steel beam under full composite action	in
$S = A_s$	area of steel beam	in^2
$D = d$	depth of steel beam	in
$C = A_{\text{ctr}}$	area of transformed concrete	in^2
$Y = Y2$	distance from top of steel beam to centroid of concrete compressive area	in

Equations 27-28: Ordinates of Reinforced Concrete Interaction Diagrams

$$x = \frac{\phi M_n}{A_g h}$$

$$y = \frac{\phi P_n}{A_g}$$

Entered as

```
X=P×M÷A÷H
```

```
Y=P×N÷A
```

Variables

$X = x$	value on x-axis	ksi
$P = \phi$	strength reduction factor	–
$M = M_n$	nominal moment	in-kips
$A = A_g$	gross area	in^2
$H = h$	effective height	in
$Y = y$	value on y-axis	ksi
$N = P_n$	nominal axial load	kips

Equation 29: Development Length of Flexural Reinforcement

$$l_d = \frac{3 d_b f_y \beta \gamma}{50\sqrt{f'_c}}$$

Entered as

```
L=3×D×Y×B×G÷50÷SQRT(C)
```

Variables

$L = l_d$	development length	in
$D = d_b$	diameter of bar	in
$Y = f_y$	yield strength of steel	lbf/in^2
$B = \beta$	epoxy coating factor	–
$G = \gamma$	size factor	–
$C = f'_c$	compressive strength of concrete	lbf/in^2

Notes The values of B (β) and G (γ) are:

$B = 1.5$	epoxy-coated, clear spacing $\leq 6d_b$
$B = 1.2$	for all other epoxy-coated bars
$B = 1.0$	uncoated bars
$G = 0.8$	no. 6 bars and smaller
$G = 1.0$	no. 7 bars and larger

Equation 30: Stability Interaction Criterion

$$\frac{f_a}{F_a} + \frac{C_{mx} f_{bx}}{\left(1 - \frac{f_a}{F'_{ex}}\right) F_{bx}} + \frac{C_{my} f_{by}}{\left(1 - \frac{f_a}{F'_{ey}}\right) F_{by}} \leq 1.0$$

Entered as

```
C=A÷B+M×R÷(1-A÷X)÷E+N×S÷(1-A÷Y)÷F
```

Variables

C	stability interaction criterion	–
$A = A_a$	allowable computed stress	kips/in^2
$B = F_a$	allowable stress	kips/in^2
$M = C_{mx}$	coefficient in interaction equation, strong axis	–
$R = f_{bx}$	computed stress, strong axis	kips/in^2
$X = F'_{ex}$	Euler stress divided by factor of safety, strong axis	kips/in^2
$E = F_{bx}$	allowable stress, strong axis	kips/in^2
$N = C_{my}$	coefficient in interaction equation, weak axis	–
$S = f_{by}$	computed stress, weak axis	kips/in^2
$Y = F'_{ey}$	Euler stress divided by factor of safety, weak axis	kips/in^2
$F = F_{by}$	allowable stress, weak axis	kips/in^2

Equation 31: Middle Ordinate

$$M = R\left(1 - \cos\frac{I}{2}\right)$$

Entered as

```
M=R×(1-COS(I÷2))
```

Variables

$M = M$	middle ordinate	ft	m
$R = R$	curve radius	ft	m
$I = I$	interior angle	deg	deg

Equation 32: Elevation at Point on Curve

$$z_x = \left(\frac{R}{2}\right)x^2 + G_1 x + z_{\text{BVC}}$$

Entered as

```
Z=R÷2×SQ(X)+G×X+B
```

Variables

$Z = z_x$	elevation at point on curve	ft
$R = R$	rate of change	ft/sta^2
$X = x$	distance from beginning of vertical curve to point	sta
$G = G_1$	initial grade	ft/sta
$B = z_{\text{BVC}}$	elevation at beginning of vertical curve	ft

Equation 33-34: Stopping Sight Distance on Crest Vertical Curve

$$S < L: \quad L = \frac{AS^2}{200\left(\sqrt{h_1} + \sqrt{h_2}\right)^2}$$

$$S > L: \quad L = 2S - \frac{200\left(\sqrt{h_1} + \sqrt{h_2}\right)^2}{A}$$

Entered as

```
L=A×SQ(S)÷200÷SQ(SQRT(G)+SQRT(H))
L=2×S-200×SQ(SQRT(G)+SQRT(H))÷A
```

Variables

$L = L$	length of curve	ft	m
$A = A$	absolute value of change in gradient	decimal	decimal
$S = S$	sight distance	ft	m
$G = h_1$	height of driver's eyes	3.5 ft	1.080 m
$H = h_2$	height of object sighted	2.0 ft	0.600 m

Equation 35-36: Stopping Sight Distance on Sag Vertical Curve

$$S < L: \quad L = \frac{AS^2}{X + 3.5S}$$

$$S > L: \quad L = 2S - \left(\frac{X + 3.5S}{A}\right)$$

Entered as

```
L=A×SQ(S)÷(X+3.5×S)
L=2×S-(X+3.5×S)÷A
```

Variables

$L = L$	length of curve	ft	m
$A = A$	change in gradient	decimal	decimal
$S = S$	sight distance	ft	m
$X = X$	adjustment factor	200	120

Notes When solving, enter 200 for X when using feet, or 120 when using meters.

Equation 37: Heavy Vehicle Adjustment Factor

$$f_{\text{HV}} = \frac{1}{1 + P_T(E_T - 1) + P_R(E_R - 1)}$$

Entered as

```
F=1÷(1+P×(T-1)+R×(E-1))
```

Variables

$F = f_{\text{HV}}$	heavy vehicle adjustment factor	–
$P = P_T$	percentage of traffic, trucks and buses	%
$R = P_R$	percentage of traffic, RVs	%
$T = E_T$	passenger car equivalent, trucks and buses	–
$E = E_R$	passenger car equivalent, RVs	–

Equation 38: Constant Acceleration

$$s = s_0 + \mathrm{v}_0 t + \frac{a_0 t^2}{2}$$

Entered as

```
S=D+V×T+A×SQ(T)÷2
```

Variables

$S = s$	distance	ft	m
$D = s_0$	initial distance	ft	m
$V = \mathrm{v}_0$	initial velocity	ft/sec	m/s
$T = t$	time	sec	s
$A = a_0$	acceleration	ft/sec^2	m/s^2

Equation 39: Average Control Delay

$$d = 0.50C\left(\frac{\left(1-\frac{g}{C}\right)^2}{1-\left(\min(1,X)\left(\frac{g}{C}\right)\right)}\right)\left(\frac{(1-P)\,f_{\mathrm{PA}}}{1-\frac{g}{C}}\right) + 900T\left((X-1)+\sqrt{(X-1)^2+\left(\frac{8kIX}{cT}\right)}\right)$$

Entered as

```
D=0.5×C×SQ(1-G÷C)÷(1-M×G÷C)×(1-P)×F÷(1-G÷C)
    +900×T×(X-1+SQRT(SQ(X-1)+8×K×I×X÷Z÷T))
```

Variables

$D = d$	average control delay	sec	s
$C = C$	cycle length	sec	s
$G = g$	effective green time	sec	s
$M = \min(1, X)$	lesser of 1 and X	–	–
$P = P$	proportion of vehicles arriving on green	–	–
$F = f_{\mathrm{PA}}$	supplemental adjustment factor for when platoon arrives during green	–	–
$T = T$	duration of analysis period	hr	h
$X = X$	volume-capacity ratio of lane group	–	–
$K = k$	incremental delay factor, dependent on controller settings	–	–
$I = I$	upstream filtering/metering adjustment factor	–	–
$Z = c$	lane group capacity	vph	vph

Equation 40: Free Flow Speed

$$\text{FFS} = S_{\text{FM}} + 0.00776\left(\frac{V_f}{f_{\text{HV}}}\right)$$

Entered as

```
F=S+0.00776×V÷H
```

Variables

$F = \text{FFS}$	free flow speed	mph	km/h
$S = S_{\text{FM}}$	mean traffic speed measured in the field	mph	km/h
$V = V_f$	observed flow rate	vph	vph
$H = f_{\text{HV}}$	heavy vehicle adjustment factor	–	–

Equation 41: Manning Equation

$$Q = \frac{K}{n}AR^{2/3}S^{1/2}$$

Entered as

```
Q=K÷N×A×R^(2÷3)×S^0.5
```

Variables

$Q = Q$	quantity of flow	ft^3/sec	m^3/s
$K = K$	constant	1.49	1.00
$N = n$	Manning roughness coefficient	–	–
$A = A$	cross-sectional area	ft^2	m^2
$R = R$	hydraulic radius	ft	m
$S = S$	slope of energy line	decimal	decimal

Notes When solving, enter 1.49 for K if using U.S. customary units, and 1 if using SI units.

Equation 42: Energy Equation (Bernoulli Equation)

$$\frac{p_2}{\gamma} + z_2 + \frac{\text{v}_2^2}{2g} + h_f = \frac{p_1}{\gamma} + z_1 + \frac{\text{v}_1^2}{2g}$$

Entered as

```
P÷W+Z+SQ(V)÷2÷G+H=O÷W+Y+SQ(U)÷2÷G
```

Variables

$P = p_2$	pressure at point 2	lbf/ft^2	Pa
$W = \gamma$	specific weight	lbf/ft^3	N/m^3
$Z = z_2$	elevation at point 2	ft	m

$V = \text{v}_2$	velocity at point 2	ft/sec	m/s
$G = g$	gravitational acceleration	32.2 ft/sec^2	9.81 m/s^2
$H = h_f$	head loss due to friction	ft	m
$O = p_1$	pressure at point 1	lbf/ft^2	Pa
$Y = z_1$	elevation at point 1	ft	m
$U = \text{v}_1$	velocity at point 1	ft/sec	m/s

Notes This equation can be used to solve problems with or without head loss. If there are no friction losses to take into account, enter zero as the value for H. For problems in real fluid flow or steady, incompressible flow in conduits and pipes, H stands for h_f.

Equation 43: Darcy-Weisbach Equation

$$h_f = f\frac{L\text{v}^2}{d2g}$$

Entered as

```
H=F×L×SQ(V)÷D÷2÷G
```

Variables

$H = h_f$	head loss due to friction	ft	m
$F = f$	Darcy friction factor	–	–
$L = L$	length	ft	m
$V = \text{v}$	velocity	ft/sec	m/s
$D = d$	diameter	ft	m
$G = g$	gravitational acceleration	32.2 ft/sec^2	9.81 m/s^2

Equation 44: Specific Speed of Pump

$$n_s = \frac{n\sqrt{Q}}{h_A^{0.75}}$$

Entered as

```
S=N×SQRT(Q)÷H^0.75
```

Variables

$S = n_s$	specific rotational speed	–
$N = n$	rotational speed	rpm
$Q = Q$	flow rate	gpm
$H = h_A$	head added by pump	ft

Equation 45: Hydraulic Radius of Trapezoidal Channel

$$R = \frac{bd\sin\theta + d^2\cos\theta}{b\sin\theta + 2d}$$

Entered as

```
R=(B×D×SIN(T)+SQ(D)×COS(T))÷(B×SIN(T)+2×D)
```

Variables

$R = R$	hydraulic radius	ft	m
$B = b$	channel width at base	ft	m
$D = d$	channel depth	ft	m
$T = \theta$	angle of channel side from horizon	deg	deg

Equations 46-47: Confined Stream in Pipe Bend

$$F_x = p_2A_2\cos\theta - p_1A_1 + \frac{\dot{m}\,(\mathrm{v}_2\cos\theta - \mathrm{v}_1)}{g_c}$$

$$F_y = \left(p_2A_2 + \frac{\dot{m}\mathrm{v}_2}{g_c}\right)\sin\theta$$

Entered as

```
X=P×A×COS(T)-O×Z+M×(V×COS(T)-U)÷G
Y=(P×A+M×V÷G)×SIN(T)
```

Variables

$X = F_x$	x-component of force	lbf	N
$O = p_1$	pressure at point 1	lbf/ft^2	Pa
$P = p_2$	pressure at point 2	lbf/ft^2	Pa
$Z = A_1$	cross-sectional area at point 1	ft^2	m^2
$A = A_2$	cross-sectional area at point 2	ft^2	m^2
$M = \dot{m}$	mass flow rate	lbm/sec	kg/s
$Y = F_y$	y-component of force	lbf	N
$U = \mathrm{v}_1$	velocity at point 1	ft/sec	m/s
$G = g_c$	gravitational constant	32.2 lbm-ft/lbf-sec^2	1
$V = \mathrm{v}_2$	velocity at point 2	ft/sec	m/s
$T = \theta$	pipe bend angle	deg	deg

Notes With U.S. units, enter 32.2 for G. With SI units, enter 1.

Equation 48: Conjugate Depths for Rectangular Channel

$$d_1 = -\tfrac{1}{2}d_2 + \sqrt{\frac{2\mathrm{v}_2^2 d_2}{g} + \frac{d_2^2}{4}}$$

$$d_2 = -\tfrac{1}{2}d_1 + \sqrt{\frac{2\mathrm{v}_1^2 d_1}{g} + \frac{d_1^2}{4}}$$

Entered as

```
C=-0.5×D+SQRT(2×SQ(V)×D÷G+SQ(D)÷4)
```

Variables

$C = d_1$	depth at point 1	ft	m
$D = d_2$	depth at point 2	ft	m
$V = \mathrm{v}_2$	velocity at point 2	ft/sec	m/s
$G = g$	gravitational acceleration	32.2 ft/sec^2	9.81 m/s^2

Equation 49: FAA Formula for Time to Concentration

$$t_c = \frac{(1.8)\,(1.1 - C)\,\sqrt{L_o}}{S^{1/3}}$$

Entered as

```
T=1.8×(1.1-C)×SQRT(L)÷S^(1÷3)
```

Variables

$T = t_c$	time to concentration	min
$C = C$	rational runoff coefficient	–
$S = S$	slope	%
$L = L_o$	length	ft

Notes When solving, enter slope as a percentage.

Equation 50: Throat Velocity in Venturi Meter

$$\mathrm{v}_2 = \left(\frac{C_\mathrm{v}}{\sqrt{1-\beta^4}}\right)\sqrt{\frac{2g\left(\rho_m-\rho\right)h}{\rho}}$$

Entered as

```
V=C÷SQRT(1-B^4)×SQRT(2×G×(M-P)×H÷P)
```

Variables

$V = \mathrm{v}_2$	throat velocity	ft/sec	m/s
$C = C_\mathrm{v}$	velocity coefficient	–	–
$B = \beta$	beta ratio (throat diameter/upstream diameter)	–	–
$G = g$	gravitational acceleration	32.2 ft/sec^2	9.81 m/s^2
$M = \rho_m$	manometer fluid density	lbm/ft^3	kg/m^3
$P = \rho$	water density	lbm/ft^3	kg/m^3
$H = h$	manometer reading	ft	m